Becoming a Place of Unrest

Becoming a Place of Unrest

Environmental Crisis and Ecophenomenological Praxis

Robert Booth

OHIO UNIVERSITY PRESS ATHENS

Ohio University Press, Athens, Ohio 45701
ohioswallow.com

Printed in the United States of America
Ohio University Press books are printed on acid-free paper ™

31 30 29 28 27 26 25 24 23 22 21   5 4 3 2 1

Library of Congress Cataloging-in-Publication Data
Names: Booth, Robert, 1983– author.
Title: Becoming a place of unrest : environmental crisis and ecophenomenological Praxis / Robert Booth.
Description: Athens : Ohio University Press, [2021] | Series: Series in continental thought ; no. 55 | Includes bibliographical references and index.
Identifiers: LCCN 2021003011 (print) | LCCN 2021003012 (ebook) | ISBN 9780821424568 (acid-free paper) | ISBN 9780821447420 (pdf)
Subjects: LCSH: Environmental sciences—Philosophy. | Nature—Effect of human beings on. | Environmental ethics. | Phenomenology.
Classification: LCC GE40 .B675 2021 (print) | LCC GE40 (ebook) | DDC 304.2/8—dc23
LC record available at https://lccn.loc.gov/2021003011
LC ebook record available at https://lccn.loc.gov/2021003012

For Wilson, Elkie, and the promise of something better

CONTENTS

PREFACE

I write this as the Himalayas become visible from India for the first time in three decades and most of the world's major cities report huge improvements in air quality (Boyle 2020). These changes reflect dramatic reductions in air pollutant concentrations from traffic and factory emissions. But none of this improvement is the result of some sudden environmentalist awakening. It stems, rather, from restrictive public health measures imposed in response to a global pandemic. It's frustratingly commonplace, therefore, to read of such changes described as "temporary," even by parties such as the United Nations Conference on Trade and Development who appear otherwise optimistic about the possibility of making substantive improvements (Hamwey 2020). For all the talk of "building back better" (Institute for Public Policy Research 2020, 14), optimism among experts that this is the beginning of some global sea change toward environmentally responsible attitudes and behaviors is in short supply.

Even setting aside the pandemic's own unwelcome results (e.g., huge increases in unrecyclable waste and greenhouse gas emissions), the global picture is worse than ever. Limiting ourselves to climate change, for instance, scientists estimate that human activities have now caused a mean global temperature increase of 1 degree Celsius over preindustrial levels. The rapid acceleration of anthropogenic emissions over the past ten years, which ensures rising temperatures to come, shows little sign of slowing. Even in the now increasingly unlikely event that we act to limit mean global temperature rises to 1.5 degrees Celsius (as per the Paris Agreement's more ambitious goal), over the next few decades, 350 million people will suffer water scarcity, 80–90 percent of coral reefs will disappear, 69 million people will experience sea flooding, and 14 percent of the world's population will be exposed to extreme heat. Much of this is already well underway (Hamwey 2020; Steffen, Crutzen, and McNeill 2007; Steffen et al. 2018; Intergovernmental Panel on Climate Change 2018).

Worse, scientists like Will Steffen now warn that a rise of even 1.5 degrees Celsius may constitute a "tipping point," where "self-reinforcing [biogeophysical] feedbacks" will ensure continued warming, even if we dramatically slash emissions (Steffen et al. 2018, 8252–54). The "Hothouse Earth" trajectory established would likely cause an even more horrifying and "uncontrollable" domino effect, resulting in global economic collapse and unprecedented fatalities due to further ocean acidification, terrestrial biodiversity loss, and land becoming too arid to support agriculture, especially in Africa (Steffen et al. 2018, 8256).

In a cruel twist, research suggests that zoonotic infections like those associated with the present pandemic will now increase in frequency and severity due to changes in migratory patterns and increasing (and increasingly risky) "human-animal" interactions linked to deforestation, food insecurity, and extreme weather all associated with human-induced climate change (Zhang 2020). Each of the above is, and will continue to be, experienced disproportionately severely by the world's poorest and most insecure populations.

This sounds shocking. Except it isn't; we aren't obviously much *shocked* by it at all. Scientific consensus about climate change is now more or less unanimous, and its headlines frequent our televisions, newspapers, and social media feeds (Powell 2019). However, for most of us, not only do terrifying factoids and the commission of ever more damning reports not translate into the behavioral and attitudinal changes apparently required, we seem to be bored of hearing about them. In the United States, for example, stories about climate change are a "palpable ratings killer" (Chris Hayes, quoted in Pope and Hertsgaard 2019), which explains why few major newspapers covered the landmark 2018 UN Intergovernmental Panel on Climate Change report warning that we have only ten years to "slash" carbon emissions before "hundreds of millions of people worldwide [will] go hungry or homeless or worse" (Pope and Hertsgaard 2019).

It may be that some are resistant to accepting climate change (and, therefore, uninterested in it), whether because of ignorance (willful or otherwise) or the influence of "cultural cognition," whereby beliefs about scientific matters reflect our political and cultural identities more than our reasoned conclusions (Kahan 2016). Likewise, as Donald Trump's withdrawal from the Paris Agreement suggests, some may perceive climate "issues" to be less important or pressing than concerns pertaining to employment, the economy, housing, food, or health care, each of which may be in tension with what crisis mitigation or aversion demands (BBC 2017). The situation might equally be attributed to uninspiring media coverage, climate change fatigue, or a general feeling of hopelessness.

I'm not convinced, however, that, individually or collectively, the foregoing is the whole story. My hunch might be supported with reference to the huge number of people who are formally committed to doing their bit to turn the present tide but do remarkably little about it at the level of everyday habit, despite having reasonable (e.g., economic) means of doing so. For every Greta Thunberg, there are thousands of others whose commitment extends to little more than switching to a renewable energy supplier or buying organic vegetables. This strange disconnect between thought and action motivates the present work.

I don't mean to imply that what follows is a treatise on the moral purity or seriousness with which one, as a reflective and self-directed agent, chooses to commit to environmental activism. I suspect, rather, that our unhelpful attitudes and behaviors largely reflect the ways that we, as situated, embodied beings, habitually (and, perhaps, unwittingly) encounter the more-than-human world at a relatively unreflective level. In this book, with the help of ecofeminists, new materialists, and, particularly, the phenomenologist Maurice Merleau-Ponty, I investigate this suspicion. I also explore how reconceiving the environmental crisis as a more fundamental matter of *praxis* (rather than a piecemeal ensemble of individual problems) might inform our attempts to overcome the aforementioned disconnect and foster more positive means of engagement with the world at large. In doing so, I aim to contribute important insights that should direct and imbue the habits of thought and action that we rightly concerned parties develop in our everyday lives. This book should be read in the spirit of optimism about the potential for impactful change to which we, as individual actors, can contribute. We surely need it.

ACKNOWLEDGMENTS

I am indebted to two anonymous reviewers and the following people, each of whom has contributed greatly to this book's development: Simon Hailwood, Jan Jobling, Simon P. James, Rachael Wiseman, Chris Cuomo, Rick Huard, Hanne Jacobs, Sally Welch, Nancy Basmajian, Mike Foden, and my ever-patient wife, Katie Booth. I cannot thank you enough.

Some material is adapted from my PhD thesis, "Environmental Crisis and Ecophenomenological Praxis" (PhD diss., University of Liverpool, 2018). Earlier versions of chapters 4 and 6 were published as "Merleau-Ponty, Correlationism, and Alterity" (*PhaenEx* 12, no. 2 [2018]: 37–58) and "Acknowledging the Place of Unrest: Tensions between Radical Reflection and the Flesh" (*Environmental Philosophy* 13, no. 1 [2016]: 57–81), respectively. I'm grateful to the editors of those journals for permitting me to reproduce that material here.

ABBREVIATIONS

Works by Maurice Merleau-Ponty:

CRO "The Child's Relation with Others." Translated by W. Cobb. In *The Primacy of Perception,* edited by J. M. Edie, 96–156. Evanston, IL: Northwestern University Press, 1964.

EM "Eye and Mind." Translated by Carleton Dallery. In *The Primacy of Perception*, edited by J. M. Edie, 159–92. Evanston, IL: Northwestern University Press, 1964.

HT *Humanism and Terror: The Communist Problem.* Translated by John O'Neill. New Brunswick, NJ: Transaction Publishers, 2000.

ILVS "Indirect Language and the Voices of Silence." In *Signs.* Translated and edited by Richard C. McLeary, 39–83. Evanston, IL: Northwestern University Press, 1964.

IPOP *In Praise of Philosophy.* Translated by J. Wild and J. M. Edie. Evanston, IL: Northwestern University Press, 1963.

N *Nature: Course Notes from the Collège de France.* Translated by Robert Vallier. Evanston, IL: Northwestern University Press, 2000.

PP *Phenomenology of Perception.* Translated by Colin Smith. London: Routledge, 1962.

PrP "The Primacy of Perception and Its Philosophical Consequences." Translated by J. M. Edie. In *The Primacy of Perception,* edited by J. M. Edie, 12–42. Evanston, IL: Northwestern University Press, 1964.

PS "The Philosopher and His Shadow." In *Signs,* translated and edited by Richard C. McLeary, 159–81. Evanston, IL: Northwestern University Press, 1964.

SB *The Structure of Behaviour.* Translated by A. L. Fisher. London: Methuen, 1965.

SNS *Sense and Non-Sense*. Translated by Hubert Dreyfus and Patricia Dreyfus. Evanston, IL: Northwestern University Press, 1964.

VI *The Visible and the Invisible*. Translated by Alphonso Lingis. Evanston, IL: Northwestern University Press, 1968.

INTRODUCTION

THE ENVIRONMENTAL CRISIS

In August 2016, the expert Working Group of the Subcommission on Quaternary Stratigraphy (WGA) voted overwhelmingly in favor of formally recognizing a new epoch called "the Anthropocene" (Carrington 2016). Their recommendation, the WGA said, was in acknowledgment of the profound and probably irreversible impact of human behaviors on geologically significant processes and conditions via intensive agriculture, urbanization, colonization, and so on. But this is not only a geological matter. The mass disruption of biogeochemical phenomena stems from the same anthropogenic causes that—partly because of those biogeochemical disruptions—continue to have profound effects on the more-than-human world at large. The supporting evidence cited by Anthropocene proponents makes for uncomfortable reading: anthropogenic climate change; widespread deforestation; ocean acidification and "spreading oceanic 'dead zones'"; desertification; ecosystem collapse; biodiversity loss; mass species extinction; and "giant swirls of plastic debris about the size of Texas in the Pacific and Atlantic oceans" (Subcommission on Quaternary Stratigraphy 2016; Crutzen 2002, 23; Steffen, Crutzen, and McNeill 2007; Bennett 2011, 120).

I raise the issue of "the Anthropocene," not because I recommend adopting the term (I don't, for reasons that will become apparent in chapter 3), but because it helpfully illustrates two aspects of the state of affairs I wish to address. The first is that the evidence cited by Anthropocene proponents is emblematic of an era of environmental crisis in which we find ourselves. Although they acknowledge its deeper roots, the WGA dates the beginning of the Anthropocene at around 1945 with a "Great Acceleration" (Steffen, Crutzen, and McNeill 2007, 617) of greenhouse gas emissions, fossil fuel and water consumption, and the volume of consumer waste deposits. While awareness of "environmental issues" has infiltrated common consciousness and public debate over the intervening years, however, this awareness hasn't led to significant mitigation of the mindset responsible for the troubling phenomena noted above, the relative situations of which worsen at an alarming rate. That isn't to say that no improvements have been made. The regulation of chlorofluorocarbons, for instance, *has* slowed anthropogenic ozone layer depletion. Neither increased awareness of ecological "issues" nor the sort of regulatory approach employed to curb chlorofluorocarbon emissions has, however, resulted in the wholesale *attitudinal* changes necessary to disrupt the gamut of problematic behaviors at root. In no significant way has broad acknowledgment of our crisis situation unseated our wasteful consumer habits, for instance, nor has it obviously undermined the attraction of economic imperatives which demand a culture of ecologically unsustainable growth. As Paul Crutzen, the atmospheric chemist credited with coining the term "Anthropocene," puts it: "The Great Acceleration is reaching criticality. Enormous, immediate challenges confront humanity over the next few decades as it attempts to pass through a bottleneck of continued population growth, excessive resource use and environmental deterioration. [Nevertheless] in most parts of the world the demand for fossil fuels overwhelms the desire to reduce greenhouse emissions" (Steffen, Crutzen, and McNeill 2007, 620). Regardless of any further claims about its "epochal" significance, if the evidence cited above is indicative of a more general state of affairs, it seems self-evident that the current state of relationships between humans and the more-than-human world is problematic in the extreme.

The second thing the Anthropocene literature foregrounds is how we typically conceive of the environmental crisis as a multitude of "issues" to be "solved" by the right kind of practical behaviors. Crutzen does exactly this when he suggests that our present crisis emphasizes "the enormity of humanity's responsibility as stewards of the Earth," a role Crutzen thinks we can satisfy if we "pioneer a modest, renewable, mindful, and less material lifestyle" (primarily by eating less meat and using public transport); increase funding for technoscience (particularly "bio-adaptive technologies"); and shift "from crusade to management, so that we can steer nature's course symbiotically instead of enslaving the formerly natural world" (Crutzen and Schwägerl, 2011). As the ecophenomenologist Ted Toadvine argues, however, by characterizing the debate *exclusively* in terms of an issue/solution schema, the more basic terms under which "environmental issues" are enframed (e.g., as matters of "excessive resource use") aren't really up for discussion (2009, 3).

THE ISSUE WITH "ISSUES"

One might be forgiven for thinking that the issue/solution schema isn't particularly problematic. If we find ourselves in a *crisis* which requires our urgent attention, then one might think that (for the time being) we *should* focus exclusively on the right kinds of practical responses.[1] But this would be a mistake for two reasons. Firstly, because, as Neil Evernden (1993, xi) suggests, understanding the environmental crisis as an *overwhelming* collection of "issues" inhibits sustained, engaged action in the wholesale manner seemingly required. Secondly, and more importantly, because the *urgency* of these calls to specifically practical action also prevents us from recognizing that "environmental issues" are "simply the visible portion of a much larger entity, most of which lies beneath the surface, beyond our daily inspection. The submerged mass constitutes the fundamental 'problem,' that domain of unspoken assumptions which legitimates, indeed even demands, the behaviour which precipitates the state of affairs we designate as 'the environmental crisis'" (Evernden 1993, xii).[2] According to Evernden, by committing our resources to tackling "environmental issues" in our present piecemeal fashion, we're doomed never to address the crisis itself, because some of its root causes remain unchallenged.

Like Evernden, I suspect that our ongoing failure to amend the behaviors and attitudes which precipitate the crisis stems largely from the ways that "environmental issues" and subsequently plausible "solutions" have hitherto been set up. If so, then any successful attempt to address them requires some investigation of the basic terms of debate. What I want to suggest in this book is that one fruitful avenue of investigation pertains to the violence already implicit in the dominant ways that we—primarily through adherence to the objectivistic onto-epistemological lens typical of the natural sciences—conceptualize more-than-human entities and our relationships to them. There can be no clear distinction between violent behaviors and their accompanying perceptual or conceptual schemes, I will suggest, because any such distinctions will fail to appreciate that our theoretical pronouncements are already intimately related to our habitual relationships with the world we investigate. Any such distinctions thus rely on the possibility of neatly distinguishing theory from praxis, which I deny. Tackling our environmental crisis wholesale, I will argue, requires attention to the (meta)theoretical apparatuses which license endemic behavioral violence, because those apparatuses already do violence to the more-than-human world through the limited ontological and epistemological terms they permit.[3] Fleshing out this claim is the main task of chapters 1, 2, and 3.

Even setting aside my specific contentions for the moment, given that our improving knowledge hasn't translated into sufficiently radical engagement with environmental "issues," Evernden's suspicions appear to warrant investigation. Since attending to our "submerged mass" appears to be a distinctly philosophical undertaking, one might think this task should fall to "environmental philosophy." However, as a matter of historical accuracy, Toadvine (2009, 4) notes that, at least in the Anglophone tradition, environmental philosophy has generally neglected this task and taken up the more superficial issue/solution schema. To understand why, one must appreciate that "philosophies of nature" have a longer history than environmental philosophy in its current guise, in which it is primarily conceived of as environmental *ethics*.

As Bruce Foltz and Robert Frodeman (2004, 2–3) note, "philosophies of nature" enjoyed a long period of historical development, shaped by predominantly teleological conceptions of the more-than-human world, prior to modernity when—doubtless partly because of the incredible predictive successes of the natural sciences—reductive, mechanistic, and value-free conceptions of the more-than-human world began to prevail. As Bruno Latour (1993) has influentially observed, the early modernist schemas of Bacon, Descartes, Newton, and Galileo, unlike those of many of their forebears, promised purification of separate realms of enquiry: an accessible world of wholly knowable natural *objects*-in-themselves on the one hand and a separable world of interrogative *subjects* on the other.

Kant inaugurated a well-known challenge to this picture by problematizing the attempt to purify reality of epistemic relations to its human investigator. One way to understand the mistake that Kant diagnoses in precritical metaphysics is as follows: by failing to appreciate that one's knowledge of "objects" cannot be entirely uncoupled from the "subjective" conceptual structures one imposes upon them, the dogmatic metaphysician unwittingly incorporates the more-than-human world into ours, mistaking the *correlational* character of the world-*for-us* for the (allegedly otherwise unknowable) world-*in-itself*. As Foltz and Frodeman note, the confines of the Kantian schema therefore dictate that specifically philosophical reflections on the world's character must be fundamentally introverted: limited to "'formal,' transcendental reflection on the possibilities and operations of our positive knowledge of nature, with the 'material' or contentful knowledge of nature restricted to the safe harbors and placid waters of the natural sciences" (2004, 2).

Although the post-Kantian picture differs radically in many respects from its predecessor, one thing remains common to both: ontology and epistemology remain relatively securely within the confines of the natural sciences. Any exploration of the "objective" character of the more-than-human world is, therefore, limited largely by the same exclusionary scientific assumptions and apparatuses which delineate the character we may discern in it. The main difference between precritical dogmatism and post-Kantian positivism concerns whether natural scientists can legitimately aim for correspondence between their findings and the "mind-independent" world. Given the confines of the Kantian schema, the positivists, insofar as they consistently reject access to the "purified" objects-in-themselves of the precritical schema, hold that they cannot.

Since environmental ethics emerged as a discipline in the 1970s against the backdrop of positivism, Toadvine argues, it was inevitable that it would retain the "Kantian division of theoretical from practical knowledge" (2009, 5). Adherence to this division would explain why environmental ethics often continues to find itself limited to the issue/solution schema, concerning itself with "purely axiological questions," rather than "raising deeper and more radical questions about the philosophical assumptions underlying our scientific and political commitments" (2009, 5). Even today, the basic terms many environmental philosophers use to identify the objects of their concern (e.g., "ecosystem" and "biodiversity") are adopted relatively uncritically from ecological science and so, given the Kantian division of labor, "predate a systematic philosophical exploration" (Foltz and Frodeman 2004, 4) of the commitments already implied by them. Perhaps more troublingly, because ontology continues to be afforded a certain deliberative priority over axiology, scientific terms also appear to dictate the shape that "solutions" (frequently understood as "moral obligations") must take. This is exactly what we see happen in Crutzen's case: his third "solution" explicitly draws its justification from the scientist Alexander von Humboldt's ecological holism. For motives similar to Crutzen's, the imperative to adopt ecologically symbiotic behaviors has become a popular sort of "solution" in the environmental ethics literature.[4]

I emphasize the development of environmental ethics within the post-Kantian context for several reasons, each of which is important in setting up the terms of the present inquiry. The first is straightforward. Following Evernden and Toadvine, I've suggested that some philosophical questioning of our basic assumptions and onto-epistemological apparatuses is required if we're to establish the wholesale engagement seemingly necessary to disrupt the behaviors responsible for our crisis situation. However, this (meta) theoretical groundwork falls outside the usual axiological scope of environmental *ethics*.

Given their near monopoly on ontological and epistemological matters, any such task would apparently fall to the natural sciences. But it's hard to see how they could carry out a sufficiently radical investigation of their own terms, given the narrow confines of science's positivistic remit within the post-Kantian context, where, as Charles S. Brown observes, "the modern enframing of nature results in a conception of nature consisting entirely of extensional properties related to one another within a causal matrix" (2003, 3). Any *violence* that such enframing might do to the more-than-human world isn't the sort of thing that can show up, and, therefore, be addressed, within a positivistic framework (Evernden 1993, 23). Furthermore, if, as I will argue, the way that we ordinarily take the more-than-human world to be *exhaustively* enframed by scientific terms is largely a persisting (but empirically invisible) *philosophical and political* relic of the problematic modernist schema championed by Descartes and Bacon, then science's background assumptions cannot uncritically constitute the boundary conditions of the radical sort of investigation I have in mind. We need to adopt a critical approach to environmental philosophy and its orientation relative to natural science if we're to attend to the necessary groundwork in earnest. That is one main task of this book.

DOGMATIC NATURALISM VS. CORRELATIONISM

The foregoing also helps highlight some other concerns that any critical environmental philosophy must navigate. One such concern relates to the relationships between realism, metaphysics, and critique.

Despite our reservations about its stronghold on epistemological and ontological matters, it should be obvious *why* environmental philosophers have sought to retain a starring role for natural science: doing so helps establish the requisite "realism" within our philosophies to address the more-than-human world *itself*. Even Kant (1996, 402/A370) subscribes to an "empirical realism" whereby the "objective" properties of existent worldly entities *really* can be known, if only as they are given a set of conceptual assumptions. Without commitment to some sort of metaphysically real world—which is pervasive in pre- and post-Kantian contexts—it's hard to see how one could be moved by the evidence cited at the outset. Why worry about the disappearing habitats of polar bears or pygmy three-toed sloths, for instance, unless one thinks that there really are some such entities whose habitats are disappearing?[5] But science does more than afford the bare possibility of "realism." Science also provides environmental philosophy with increasingly sophisticated and fine-grained observations that are *essential* to understanding the contours of our crisis situation (how else could we begin to grasp the harm in destabilizing forest ecosystems or phosphorus cycles, for example?) and some possible means of addressing it (via resource allocation models, and so on).

Nevertheless, respecting science's valuable contributions cannot require commitment to any sort of dogmatic metaphysical naturalism under which the ontological reality uncovered by the natural sciences is taken to straightforwardly correspond to the mind-independent world. This position is problematic for reasons that we've already seen Kant identify. Kant suggests that dogmatic metaphysicians misrepresent more-than-human entities by projecting correlational epistemic properties—which are nonaccidentally related to the kinds of perceivers we are—*onto* "things-in-themselves." Theirs is more or less the same mistake identified by the German biologist Jakob von Uexküll (1957). Uexküll suggests that even human organisms tend to mistake their *Umwelt* (or *environment*)—complete with biosemiotic signs specifically related to the biological capacities that organism possesses—for the *Umgebung,* or (allegedly "noumenal") reality *itself.*[6] Uexküll contends that for ticks, for example, space and time exist only relative to the chemical stimulus of butyric acid, which is straightforwardly absent from human *Umwelten.* Thus, the overall problem with taking natural science (suitably conceived: Descartes's physics is not Bacon's empiricism) to provide a straightforward window onto mind-independent reality is that it promises what Hilary Putnam calls "the metaphysical objectivity of the God's Eye view" (1981, 55), but achieves it only by subsuming more-than-human reality under our own—markedly human—conceptual schemes.[7]

In one sense, by forcing us to address the conceptual sediment that infuses our accounts of reality, Kant's Copernican Revolution represents significant improvement in our potential to reduce the ontological and epistemological violence done to the more-than-human world. If our observations of phenomena are never passive, desituated, nor epistemically innocent, as Kant and his descendants suggest, then we should surely be alive to our contributions to those phenomena in a manner that speaks against our violent appropriation of them. One might reasonably understand environmental theorists like the ecofeminists we will encounter in chapter 2 to be engaged in a related project of critical self-reflexivity through their attempts to weed out some problematic conceptual assumptions implicit in the ways that we typically enframe the more-than-human world. If, as ecofeminists suggest, the violence done in callous habitat depletion is intimately related to the violence done in "naturalizing" *our* perspectives on it—perhaps even in characterizing rainforests *as* "depleted resources," for instance—then we must presumably heed the Kantian imperative for critical self-reflexivity in any attempt to attend to our submerged mass.

In another sense, however, the introverted focus Kant forces in philosophy makes it difficult to see how philosophy can evade what Quentin Meillassoux calls the "correlationist circle" (2008, 5). In this circle, correlational terms like those identified by Kant are absolutized—in at least the epistemological sense—such that perceptual or conceptual access to nonhuman things *themselves* becomes impossible. In the correlationist model (which I explore in chapter 4), one effectively abandons any attempt to address the more-than-human-world on its own terms. The correlationist is, instead, led to make claims like the following: " 'Objects' do not exist independently of conceptual schemes. *We* cut up the world into objects when we introduce one or another scheme of description" (Putnam 1981, 45). Under correlationism, the necessity of doing pervasive violence to the world through imposition of our conceptual schemes becomes something we simply (perhaps begrudgingly) accept.

Note that one reason Putnam speaks of *multiple* schemes of description is in recognition of the multitude of correlational intermediaries between "subject" and "object" explored since Kant. For Michel Foucault or Judith Butler, this might be ideological discourse; for Ludwig Wittgenstein or Jacques Derrida, this might be language; and so on. But what is common to all correlationist philosophies, according to Meillassoux, is their totalitarianism. By moving from a reality carved up by "a Mythical mind" (in Kant) to multiple "social" accounts of correlational mediation (post-Kant), Latour argues that investigative subjects remain "locked not only into the prison of their own categories, but into that of their social groups as well" (1999, 6–7). From this limited purview, of which the environmental philosophers Steven Vogel (2002; 2015) and William Cronon (1995) are arguably key proponents, "nature" (i.e., nonhuman reality) ends up being a "profoundly human construction" (Cronon 1995, 25). This is because, while neither Cronon nor Vogel denies the existence of a metaphysically real world beyond our practical involvements with it, they nevertheless deny any possibility of access to it independently of mediation by reified human concepts (including "nature" and "environment"), culture, practices, and so on (Vogel 2002, 29–32).

Vogel's case is complicated by his contention that we (literally) construct our environments through physical labor rather than the projection of conceptual schemes. For him, "subject and object come to be what they are through practice, not the other way around" (Vogel 2015, 123). However, Vogel (2015, 14, 44–45, 123, 143) also spends a great deal of time emphasizing the capitalist purview crystallized within the web of practices that shape the world and our knowledge of it, which, according to his own logic, remains equally inseparable from how we become and understand ourselves. Thus, there is an important neo-Hegelian sense in which Vogel ontologically absolutizes the terms of the correlation, which prevents him from getting any closer to a genuinely more-than-human world on *its own terms* (i.e., independent of those imposed via value-laden capitalistic practices). The world here just is "the world we have built," and "the 'matter' of which materialism speaks is always a matter that we have built" (Vogel 2015, 73, 90). Likewise, other than providing relatively stable structural conditions for construction (the "forces of air and gravity, of heat and light, of decay and oxidation and time"), nonhuman "artifacts" retain predictably little ontological resistance or impact upon our epistemic processes beyond the minimal capacity for disruption experienced through the baker's or engineer's inability to *wholly* predict the behaviors of yeast microbes or shopping malls (2015, 112–15).[8] Insofar as they lack the linguistic sophistication to literally "speak," Vogel also denies nonhuman entities any independent capacity to correct *our* plastic and utilitarian constructions of them (see chapter 6).

Why is this significant for our present task? I've suggested that one underlying cause of the environmental crisis is our imposition of certain kinds of ontological and epistemological schema onto the more-than-human world. It's at least plausible, and therefore worthy of investigation, that the behavioral violence enacted in aggressive deforestation for palm oil cultivation, for instance, is nonaccidentally related to the violence done at a metatheoretical level through the ways we conceptualize the terms of debate. If the bounds of the correlationist circle are as inescapable as they appear, however, then we face the problem that situated human "subjects" are substantively *cleaved* from the nonhuman "objects" of their concern (rainforests, orangutans, biodiversity, etc.) and so any attempt to address these entities *themselves* appears futile.[9] If environmental philosophy is doomed to forever conceptualize—or, as Cronon and Vogel contend, *construct*—the more-than-human world from within our increasingly narrow, anthropocentric prisons, then any hope of better addressing it on its own terms appears to be lost.

Of course, even from within a correlationist or constructivist purview, environmental philosophy isn't without merit. It can remain self-reflexive about its political and philosophical commitments in the praiseworthy sense that those schemas and "practices that *know themselves as such,* that acknowledge their own social and transformative character, are to be preferred . . . over those that do not" (Vogel 2002, 35). By acknowledging the epistemic centrality of one's specific human situation in revealing more-than-human phenomena, theory might ostensibly therefore remain anthropocentric in only a benign ontological sense rather than the more obviously pernicious ethical sense where human needs become central to and sufficient for guiding human-nonhuman relationships.[10] But, if all epistemic access to the more-than-human world is socially constructed, then philosophy's acknowledgment of its inherent ontological anthropocentrism or anthropomorphism is as far as it can stretch in its commitment to mitigate onto-epistemological violence. Environmental theory would remain irredeemably anthropocentric in the potentially dangerous ontological sense that, even in critique, it couldn't penetrate the value-laden terms under which we allegedly construct "nature" in the first place. There could be no possibility of correction from *outside* the correlationist circle. Thus, if the more fundamental sort of violence we're investigating can be understood, in part, as the failure or refusal to engage with the more-than-human world on its own terms—which is what the intuitive attraction of naturalistic realism in an environmental context suggests—then correlationists and constructivists are set against addressing this sort of violence *even in principle.*

Some go further. Timothy Morton, for instance, claims that if "objectivity" is reduced to "*objectivity for us*"—where "us" denotes human beings, as Putnam concedes—then environmental theory remains irredeemably geared *toward* violence against nonhuman entities (Morton 2011; Putnam 1981, 55). Morton makes this charge for reasons similar to Uexküll's: because in the correlationist circle—*even in acts of critique*—nonhuman "things" show up *for us* only insofar as they relate to *our* interests as manifested in human concepts and language. Even if one finds Morton's claim too strong, by effectively reducing nonhuman entities to our representations or constructions of them, correlationism nevertheless appears to introvert critical environmental philosophy in a manner relatively straightforwardly incompatible with more-than-human *alterity*. If the cost of taking seriously our active contribution to the objects of our concern is that we ultimately legitimate conceptual violence against them, then, given our current project, the cost is surely too much.

The preceding discussion is too coarsely grained to do justice to the issues and approaches it touches upon. I will revisit some of them in what follows. Nevertheless, regardless of its simplicity, the foregoing should serve useful to position the argument that will be advanced in this book. To summarize:

1. Philosophers must investigate the terms by which the environmental crisis is enframed if we're to address it in earnest.
2. This task requires critical assessment of the assumptions, apparatuses, and commitments involved in the natural scientific ontologies and epistemologies which underpin our understanding of environmental "issues."
3. Since this critical assessment is a distinctly philosophical undertaking, the focus of environmental philosophy must also be a critical one.
4. Critical environmental philosophy also needs to avoid the pitfalls of dogmatic naturalism and correlationism, each of which risks doing undue violence to the more-than-human world by implicitly or explicitly subsuming it under anthropocentric terms (i.e., as it shows up for us only relative to *our* concerns, concepts, power relations, language, and so on).

AN ECOPHENOMENOLOGICAL ALTERNATIVE

In what follows, I employ *ecophenomenological* tools to attend to the task outlined above. Resituating environmental philosophy in the vein licensed by Maurice Merleau-Ponty's work, I will argue, allows it to resist uncritically adopting reductive naturalist ontologies and epistemologies. Moreover, this sort of ecophenomenology may also chart the requisite course between dogmatic realism and correlationism to address the more-than-human world on its own, rich, terms. Note, however that I say, "*licensed by* Merleau-Ponty's work." His later ontological turn notwithstanding, I believe that the ecophenomenological approach advocated in this book chimes with Merleau-Ponty's most important philosophical commitments. Nevertheless,

I'm more interested in whether this approach may help to disrupt the violence of our submerged mass than whether it is an entirely faithful rendition of Merleau-Ponty's philosophy. I will, therefore, hereafter refer to it being "Merleau-Pont*ian*."

According to the Merleau-Pontian model I advocate, our respective pitfalls may be avoided via attention to the epistemic salience of one's *embodiment* to the world one inhabits.[11] A reconstruction of my basic argument is as follows: since all ontology and epistemology are derived from a perceptual horizon which, given the fundamental subject-object ambiguity of one's embodiment, admits of "subject" or "object" poles as only secondary abstractions, then—contrary to Latour's (1993, 8–9) accusations—the ecophenomenologist may reject the terms of the modernist schema. Doing so means that the Merleau-Pontian isn't left with a choice between dogmatic naturalistic objectivity and correlational objectivity-for-us (conceived naturalistically or otherwise). Why? Because, as embodied, one is ontologically continuous with the more-than-human world investigated and so isn't denied unmediated access to it in the manner of the correlationist "*subject*" (versus correlationism). But also because ecophenomenologists deny that accessing the world on its own terms requires recourse to "mind-independent *objects*"—the sloth, polar bear, or forest ecosystem "in-themselves"—devoid of conceptual sediment or active structuring on our part (versus dogmatic naturalism).

To elaborate, it's important to realize that both dogmatic naturalist and correlationist alternatives "share a common assumption, namely, that nature in its own right lacks a sense or meaning that is open to human understanding" (Toadvine 2009, 15). The former generally denies that "meaning" exists as a basic facet of the nonhuman world because it cannot obviously be identified as a positive property-in-itself. The latter at least denies that nonhuman entities have meanings of their own that can be made available to us outside of our various conceptual schemes or social practices. This is one key reason why, for Kant and many of his environmentally minded descendants (e.g., Wilson 1998), recognition of the correlation doesn't disrupt naturalism's stranglehold on ontology and epistemology, but rather results in resituating science positivistically. In making the claims they do, however, both dogmatic naturalists and correlationists tend to license only a limited and plastic model of the more-than-human world. Both also risk making meaning, culture, and agency the preserve of *human* subjects in a suspiciously anthropocentric manner.[12]

For the Merleau-Pontian ecophenomenologist, however, reality is expressed ambiguously *between* body-subject and world, and so the more-than-human world can express its own, rich, meanings with, or through, human body-subjects. The Merleau-Pontian can, therefore, resist fundamental commitment to naturalism's reductive ontological and epistemological terms. Moreover, expressive truth becomes possible by rejecting the modernist separation of subject from object. There is, therefore, no need for the Merleau-Pontian to speak of "filtration" between poles, so any charge of a pejorative anthropocentrism derived from the apparent necessity of correlational mediation will also be overdrawn. Importantly, however, dissolving a difference in kind needn't make the body-subject and more-than-human world *the same thing*. Epistemic contact with the more-than-human world itself does not, therefore, grant the Merleau-Pontian infallible acquaintance with its meanings. Neither must the Merleau-Pontians' claim that *theirs* are the only meanings truly available for expression. Thus, Merleau-Pontian ecophenomenology finds space for both robust *critical self-reflexivity* (as is demanded post-Kant), and radical *openness* to the more-than-human world on its own, rich, terms (embracing a realism seemingly impossible post-Kant).

Since satisfying both demands appears essential for a positive and rigorous reconfiguring of the entities that environmental philosophers address, it should be clear why Merleau-Pontian ecophenomenology provides a useful toolkit for our present task. Much more must be said about all of this, however, if a convincing case is to be made for asking philosophers to reconceive the environmental crisis ecophenomenologically. Since providing the details of this case is the task of the present book, however, I resist the urge to do so here. I shall instead signpost the main contours of my argument via an outline of the following chapters.

CHAPTER OUTLINE

One main task of chapters 2 and 3 is to explore what is so problematic about our default naturalism in ontological and epistemological matters. In chapter 1, I draw principally from Merleau-Ponty's *Structure of Behaviour* and *Phenomenology of Perception* to argue that, by granting ontological status only to positive objects-in-themselves (including reified relations and processes such as ecosystems or carbon cycles-in-themselves), naturalism neglects the diacritical relations that provide the necessary perceptual background for those "objects" to present themselves with the character they do.[13] By neglecting these relations, the *objectivistic* remit of natural science affords a limited and phenomenologically suspect model of the more-than-human world. Since, I argue, the ultimate basis of ontology and epistemology is our phenomenological openings onto the world, the latter charge undermines the assumed philosophical legitimacy of "objective thought" (*PP*, 71) that underpins naturalism. Moreover, since "objective thought" deals solely in reflective abstractions, it also risks distorting the character of the phenomena that natural scientists seek to address. If, as I have suggested, science cannot itself attend to these failings, then a "return to the [more-than-human] things themselves . . . is from the start a foreswearing of science" (*PP*, viii).

I then outline key features of a Merleau-Pontian alternative which takes seriously that one's ordinary modes of bodily comportment into the world form part of the diacritical background which affords the basic meanings that the more-than-human world permits. Since it is also only *as* one's body that one may *grasp* the world's own meanings, one's body also affords unmediated contact with that world, complete with a richer and more profound set of (e.g., axiological) meanings than might otherwise be accommodated. Nevertheless, since Merleau-Ponty argues that the meanings the world affords at the level of consciousness rely on a preconscious "body schema" subject to habitual sediment, the Merleau-Pontian ecophenomenologists must remain radically reflective about their own sedimented intentional (e.g., ideological, political, and economic) commitments. Failure to recognize our intentional contributions to the world's physiognomy is part of the reason why, I will suggest, even as environmental philosophers, we unwittingly enframe the more-than-human world under the implicitly instrumentalist terms of technoscience.

Understood in the Merleau-Pontian context I suggest, the main problem in trying to discern the world's *true* ontological character is that body-subjects are embedded within the world they investigate, and so cannot enact the sort of transcendent or transcendental reflection which naturalism—as a form of objective thought—appears to require.[14] The major lesson here is that, for the Merleau-Pontian, the error involved in our objectivistic appropriation of the more-than-human world is more-than-cognitive; it is one of *praxis*. It pertains to our basic, *bodily*, ways of navigating, utilizing, and relating to the world; ways which are always partly the product of historico-cultural sediment. Since a full transcendent or transcendental reduction is impossible, this means that an appropriate approach to *addressing* problematic accounts of that world must be one of radically reflective praxis. Merleau-Pontian ecophenomenology contributes to our renewed investigation of the more-than-human world by setting itself up, more or less, *as* a certain kind of critical social praxis. The critical self-reflexivity involved in such a praxis is related to that which is of merit in Vogel's account. Unlike Vogel's environmental philosophy, however, Merleau-Pontian ecophenomenology benefits from being able to understand "seeing better" partly in terms of the accuracy with which one expresses the meanings made available for expression by the more-than-human world *itself*.

In chapter 2, I explore the powerful contributions that ecofeminism may offer ecophenomenology by providing a better account of some of the problematic *political* sediment of our submerged mass. Val Plumwood's ecofeminism, I argue, contributes to our capacity for radical reflection by highlighting that objectivistic schemas (like those commonly advanced by natural scientists) rely on a *colonial* logic. The issue here isn't only that subject and object are misleadingly set apart in kind, but also that their separation results partly from a *value* dualism, according to which "objects" are hierarchically inferior to "subjects." Thus, Plumwood allows us to see that if we're to address our environmental crisis then we must disrupt the implicitly dualistic terms under which natural science operates because these are set *violently* against nonhierarchical and nonexploitative engagement with nonhuman entities (and other Others) from the outset.

While Plumwood's basic argument is compelling, I argue that its salient features should be reconfigured ecophenomenologically. The main reason for my recommendation is because, by situating her approach naturalistically, Plumwood courts a curious disembodiment within theory. By failing to fully appreciate the intentional salience of embodiment, Plumwood leaves herself insufficiently open to the radical alterity that may imbue the ways that nonhuman animals, for instance, may differentially experience the world's meanings. Furthermore, by failing to appreciate the extent to which dualistic logic saturates the background of our ontological predispositions, Plumwood also misconstrues how best to unseat dualistic praxis.

In chapter 3, I take up the (primarily critical) insights of the previous chapters to explore how Merleau-Pontian ecophenomenology, bolstered with ecofeminist and new materialist insights, might help us perceive the more-than-human world *better*. I do this primarily by considering some provisional ways in which the praxis I defend might foreground subject/object, mind/nature, and mind/body continuities as correctives to our default naturalism's illicit dualism. I argue that naturalism's objectivistic assumptions render it incompatible with recent insights about the intentional meanings plausibly expressed by beetles and forests, and the quasi agency possessed by waste plastic. These incompatibilities force us to recognize that the epistemological and ontological commitments involved in taking the "objects" enframed by naturalism to exhaust the world—and, therefore, form *the* basis of our understanding of the "environmental crisis"—conceal a drive to mastery as part of their diacritical background. This is because they remain inattentive to the partial openings from which they are derived and the intentional commitments (e.g., the "masculine" political impetus to reduce more-than-human entities *to* resources) which infuse the background terms of engagement.

Engaging with the more-than-human world on its own terms, I argue, requires a philosophy equipped to recognize the mindlike properties that entities may possess. But it also requires recognizing that the metaphysical presupposition of a wholly knowable realm of objects-in-themselves (whether dogmatic or positivistic) is unfit for purpose. This is because the more-than-human world cannot be unproblematically reduced to a collection of determinate "objects" or properties-in-themselves. But it is also because concerned epistemic "subjects" cannot be purified of the multitude of ostensibly "objective" properties which delineate the intentional character of the objects they grasp. Not only is this because things emerge for us relative to the specific demands of our motor intentions (as Merleau-Ponty shows), but also because (as Jane Bennett argues) we cannot purify epistemic "subjects" of the contribution provided by the foods they consume, and so on.

Philosophy must acknowledge that our ontologies and epistemologies are nonaccidentally related to the irreducibly embodied modes of engagement under which they are taken up. I therefore follow Merleau-Ponty in departing from reductive naturalism and propose a new onto-epistemological norm: the "view-from-everywhere," which, as the sum of true expressions of the world, may encompass the intentional meanings expressed by divergent more-than-human entities. Taking up a view-from-everywhere, I argue, allows us to appreciate that entities can be taken up under multiple legitimate schemes of description that do not individually exhaust the meanings truly available for expression (some of which, given our limited kinds of embodied openings, will be beyond our ken). The descriptions that atmospheric chemists offer about carbon dioxide emissions, for example, may be *true*, but only granted certain regional concessions and cannot, therefore, exhaust the phenomena they describe. By embracing the more positive onto-epistemological norm required by ecophenomenological analysis, I argue, we stand a better chance of addressing the submerged mass of our environmental crisis. This is because the norm I suggest is geared away from doing undue violence to more-than-human alterity from the outset. It cannot, after all, require us to *reduce* a forest to "carbon deposits," an "ecosystem," or certain "ecosystem services." I end the chapter by discussing how we might resist objectifying habits by inculcating a praxis of radical reflection.

If chapters 1 to 3 focus broadly on Merleau-Pontian ecophenomenology's rejection of naturalism, then chapters 4 to 6 concern its rejection of correlationism. In chapter 4, I address the claim that even a radically reflective ecophenomenology is incompatible with addressing the more-than-human world *itself* in the manner required by a view-from-*everywhere*. Proponents of this charge hold that, by subsuming more-than-human entities into a human intentional horizon, phenomenology remains correlationist in the problematic (ontologically and, subsequently, ethically) anthropocentric sense already discussed. I dispute this claim for reasons already mooted. Firstly, for the Merleau-Pontian, one is never straightforwardly a human "subject" cleaved from the world-in-itself in the requisite manner for any serious correlationist charge to hold. Secondly, the critical focus of Merleau-Pontian ecophenomenology *better* equips it to resist conceptual violence against the more-than-human world than its allegedly anticorrelationist rivals. Thus, I argue, there is no in-principle incompatibility between my praxis and the goal I set for it.

In chapter 5 I discuss a narrower version of the correlationist charge which holds that, although there may be no in-principle contradiction between ecophenomenology and an anticolonial rethinking of our onto-epistemological assumptions, ecophenomenology is problematically predisposed to colonialism via its inherent *androcentrism*. This issue arises for reasons similar to the ones explored by Plumwood: in patriarchal or dualistic societies, our conceptual and perceptual schemes are dominated by problematic discursive norms which don't present themselves straightforwardly to consciousness. The main charge here is that phenomenology is prone to mistake its *androcentric* intentional heritage (e.g., by making objectification a transcendental condition of experience in Husserl's case) for the expression of the world's nondiscursive meanings; an error that, because of its introverted focus on lived-experience, ecophenomenology appears poorly equipped to uncover. Whilst it is true, I argue, that the Merleau-Pontian must engage with the more-than-human world's own nondiscursive grounds, it is false to think that reflectively uncovering them is a solitary endeavor. Rather, it is recognition of the tensions between one's transcendence and immanence that ultimately opens the Merleau-Pontian to fruitful *intersubjective* means of correction as a matter of course. Thus, ecophenomenology's potential for nonandrocentrism, like its capacity to avoid a pernicious ontological-cum-ethical anthropocentrism, derives from its rejection of the terms of the modernist schema and its subsequent recognition that a full transcendent or transcendental reduction is unobtainable.

Nevertheless, it might be objected that our view-from-everywhere doesn't seem to be straightforwardly amenable to correction by nonhuman entities *themselves* and so ecophenomenology harbors a residual tendency toward the more troubling kind of ontological anthropocentrism which limits our openness to the ethical demands or ontologically real meanings of more-than-human others. In chapter 6, I discuss Merleau-Ponty's later ontological turn, which has the potential to address this sort of issue by making all expressions constitutive of the world's own "flesh." I argue, however, that any subject-object reversibility licensed by the flesh is dialectically uninformative and may hamper the critical self-reflexivity which provides much of ecophenomenology's merit. One reason for its failings is that *this* appeal to the flesh requires a fundamental flattening of ontology that ends up obscuring or naturalizing the specific terms of the phenomenological opening under which it was flattened. Thus, such an appeal threatens openness to alterity by making the Other's unfamiliarity too familiar in a manner also seen in speculative realist metaphysics, and not entirely alien to the one seen in objective thought.

In the concluding section, I draw the strands of my argument together. One might be disappointed by the failure of a turn to the flesh because it leaves ecophenomenology in the apparently paradoxical position of conceding its own (fairly strong) ontological anthropocentrism while also claiming that its value lies in facilitating less anthropocentric accounts (in intimately related ontological *and* ethical senses) of the more-than-human world than its objectivistic counterparts. Nevertheless, I think that the Merleau-Pontian can overcome the apparent performative contradiction involved. This is because, as I have suggested, as a body-subject, one is never only, nor entirely, "human" in the requisite manner for one's own ontological anthropocentrism to preclude access to the meanings expressed by Others with whom one is connatural. Furthermore, by accepting the partiality of one's opening onto the world's meanings, one resists leveling down the Other's alterity in the manner one risks by taking up a fundamental ontology. Thus, by taking seriously one's asymmetric (or *chiasmic*) opening onto the more-than-human world (and Others therein), the

Merleau-Pontian may chart the requisite course between critical self-reflexivity and openness to alterity to resist violently appropriating the more-than-human world as a matter of habit.[15]

SCOPE

I will end this introduction by saying a few words about the scope of my project. On the one hand, the task I assign myself is ambitious: to undertake an ecophenomenological rethinking of the ways that we understand and relate to the more-than-human world. On the other hand, however, my project is relatively modest. The metatheoretical groundwork I have in mind, for instance, isn't even a completely novel one. Related projects have been attempted, with mixed results, by critical environmental theorists including deep ecologists (e.g., Arne Naess, Warwick Fox), social ecologists (e.g., Murray Bookchin), ecofeminists (e.g., Val Plumwood, Carolyn Merchant), and ecophenomenologists, some of whom I have drawn upon in this introduction (e.g., Ted Toadvine, Bryan E. Bannon, Neil Evernden, David Abram). While I cannot address their approaches individually, it might be useful to say something about why this kind of Merleau-Pontian approach may be better equipped for the task in hand than other kinds of critical environmental theory.

Existing ecophenomenological approaches foreground one common shortcoming among critical environmental theorists insofar as they fail to engage sufficiently with (eco)feminist analyses and so tend to overplay our openness to (more-than-human) alterity. As I will later illustrate with reference to alternative Merleau-Pontian ecophenomenologies, this is often because they misconstrue the extent of epistemic openness facilitated by one's embodiment. Social ecologists like Bookchin make a related error by "effectively subsuming nature under mind" (Cook 2011, 135) in the Hegelian sense. As we shall see, ecofeminist approaches exemplify another shortcoming insofar as they often fail to address the epistemic salience of one's embodiment and so are apt to either overwrite alterity by situating their approaches naturalistically, or else court a problematic correlationism. Similarly, deep ecologists (e.g., Callicott 1989; Fox 1990), for their part, often base their critiques of mainstream environmental philosophy on data appropriated relatively uncritically from ecological science and so inherit its problematic presuppositions. I contend that the approach I promote strikes a better balance between openness and critical self-reflexivity for the task in hand than do the available alternatives.

My project is also modest in terms of the things it doesn't set out to do. While my arguments for taking up a radically reflective ecophenomenological praxis may seem plausible if one thinks that we should address our fundamentally problematic relationships with the more-than-human world, I will not supply any metaethical arguments which would tie the reader into an obligation to care about doing so. Like Edward S. Casey, I take it that "an ethics of the environment must begin with the sheer and simple fact of being struck by something wrong happening in the environment" (2001, 2). There seem, to my mind, to be meanings available for expression *within* the melting ice caps and polluted oceans which provide irreducibly phenomenological reasons to do something about them. At least as a provisional hypothesis, this seems to me to be unavoidably true. But my contentions may not stir the unflinching technocrat who insists they can see no such thing and refuses to engage with even the basic demands of my project.

I also don't consider it to be within the scope of this book to address in detail which maxims or normative principles might be legitimately drawn from the praxis I espouse. This isn't because I think that no such guidance could arise out of the metatheoretical groundwork I undertake, but because, as a situated theorist myself, I must take that situation sufficiently seriously. While meaningful work can surely be done in an ecophenomenological context concerning practice-focused means of addressing anthropogenic climate change, food scarcity, biodiversity loss, and so on, any such work must be subject to an intersubjective hyperdialectic which takes seriously the fecundities between relevant disciplines (few of which I have been party to). Advancing an ethic or policy framework is not the sort of task one can plausibly achieve from one's armchair.

Nevertheless, these admissions don't make the work I'm doing so modest as to be trivial. My aim in this book is to do some of the essential groundwork for us to attend to the "submerged mass" of our environmental crisis. Thus, even though this groundwork may, at times, appear so abstract as to be removed from the sphere of practical activism, its value goes beyond any supposed environmental theory/practice distinction. Although I'm wary of his wider ontological and epistemological commitments, John Baird Callicott is worth quoting at length on why this is the case:

> All environmentalists should be activists, but activism can take a variety of forms. The way that environmental philosophers can be the most effective environmental activists is by doing environmental

philosophy. Of course, not everyone can be or wants or needs to be an environmental philosopher. Those who are not can undertake direct environmental action in other ways. My point is that environmental philosophers should not feel compelled to stop thinking, talking, and writing about environmental ethics, and go *do* something about it instead—because talk is cheap and action is dear. In thinking, talking, and writing about environmental ethics, environmental philosophers already have their shoulders to the wheel, helping to reconfigure the prevailing cultural worldview and thus helping to push general practice in the direction of environmental responsibility. (1999, 43)

CHAPTER ONE

PERCEPTION AND UNREST

> Man, as opposed to a pebble which is what it is, is defined as a place of unrest *(Unruhe)*, a constant effort to get back to himself, and consequently by his refusal to limit himself to one or another of his determinations.
>
> —*SNS*, 55–56

In my introduction, I mooted reasons for environmental philosophers to be wary of uncritically adopting epistemological and ontological assumptions from the natural sciences. In this chapter, I lay the foundations for an ecophenomenological investigation of the "submerged mass" of our crisis. I first identify and problematize some of the objectivistic prejudices that pervade science's underpinning naturalistic metaphysics. Drawing principally from Merleau-Ponty's earlier work, I argue that the "objective thought" which underpins naturalism is concerned primarily with reflective abstractions rather than phenomena-as-given. While this admission needn't be problematic per se, I follow Merleau-Ponty in suggesting that naturalism's basic onto-epistemological stipulations distort the phenomena upon which they are based.[1] The problem Merleau-Ponty identifies with objective thought is that it conflates the determinate Gestalt-objects in which perception concludes with the basic constituents of ontological reality. Over the next three chapters I take Merleau-Ponty's argument further by suggesting that the problematic ways we enframe more-than-human entities under the influence of naturalism can be traced back, in part, to the mistaken reification of "objects" upon which naturalism is based. This chapter will provide basic resources for doing so.

My central contention is that any attempt to address more-than-human entities on their own terms must take seriously the coconstitutive perceptual horizons through which they are always uncovered. Any such attempt must suspend foundational commitment to naturalism's objectivistic assumptions and attempt to go "back to the things themselves" (Husserl 2001, 168). I argue that progress in this project can best be facilitated through what Merleau-Ponty calls "radical reflection" (*PP*, xiv)—a sort of second-level reflection on the interplay between prereflective and reflective consciousnesses—when trying to glean the perceptual reality that underwrites our reflective accounts. Radical reflection may help foreground some of the background assumptions and apparatuses required for a situated body-subject to grasp a given more-than-human "object" as one does. It may also thereby help shed some light on our "submerged mass" by highlighting where our objectivistic *mis*conceptions or *unwarranted* assumptions license behavioral violence against any entities subsequently uncovered.

Although, given one's intractable inherence within the world investigated, Merleau-Ponty is surely right to claim that the process of radical reflection is incompletable, I argue that recognition of this "failure" is key to better engagement with the more-than-human world. This is because, since a fully transcendent or transcendental account is impossible, the committed Merleau-Pontian is tied to perpetual radical reflection as a matter of praxis. The Merleau-Pontian ecophenomenologist may, therefore, retain the robust critical self-reflexivity that Kant's recognition of the correlation demands, without taking up the objectivistic Kantian baggage which motivates the "dialectic of the epistemological subject and the scientific object," and which results in "making subject and object into inseparable correlatives" (*SB*, 199; 201).

1.1 OBJECTIVE THOUGHT

I begin with Merleau-Ponty's critique of "objective thought" thought—so called because of the allegedly foundational "objects" it relies upon—latent within both dominant philosophical accounts of perception: empiricism and intellectualism.[2] Although it might seem strange to begin a critique of naturalism by evaluating models of perception, Merleau-Ponty provides reason to believe that an insular dialectic exists between the misleading accounts of perception relied upon by natural scientists and the ontological and epistemological claims they make. Given that our overriding concern is to address the assumptions and commitments that delineate scientific characterizations of the more-than-human world, I will focus on Merleau-Ponty's critique of empiricism. I do so because the ontological and epistemological terms that environmental philosophers inherit derive largely from natural sciences (e.g., ecological science, atmospheric chemistry, or evolutionary biology), which are empirical in method.

1.1.1 Empiricism

Merleau-Ponty's critique of empiricism focuses on its mistaken reliance on "sensations" as the constituent atoms of perception. By "sensation," Merleau-Ponty means what we might now call qualia: discrete, nonintentional, internal phenomena, from which perceptual experience is purportedly constructed. According to the empirical account, in perceiving a telephone box, an epistemic subject is more or less directly acquainted with the self-contained sensation of "redness" in consciousness. But sensations aren't limited to simple scenarios. Sensations (or some other kind of "pure impression") are also ordinarily taken to be the source of the atmospheric chemist's observations about carbon dioxide emissions, for instance, albeit via specialized measuring apparatuses and subsequently contextualized within a complex theoretical framework. This is because, as Thomas S. Kuhn emphasizes, in attempting to discern the world's basic ontological character, the "scientist can have no recourse above or beyond what he sees with his eyes and instruments" (1970, 114).

However, despite our "commonsense" conviction in sensations, Merleau-Ponty argues that we experience no such phenomena. Furthermore, since we treat sensations as perception's fundamental constituents, "once introduced, the notion of sensation distorts any analysis of perception" (*PP*, 13). Thus, if the empirical account of perception is as intimately related to naturalistic metaphysics as Merleau-Ponty suggests, any subsequent ontological and epistemological stipulations that rely on it will be similarly distorted.

Reflection upon perceptual experiences suggests that we aren't directly acquainted with such *discrete* phenomena, with *constant and self-contained* qualities, because perceptual qualities are determined via their diacritical relationships within a perceptual field. Even a white patch, Merleau-Ponty argues, possesses its "brightness" and "shape" only relative to (the perceived color of) its particular background, which also codetermines areas "belonging" to it, and so on (*PP*, 3–4). This diacritical codeterminacy becomes most obvious in optical "illusions." In Edward Adelson's "Checkershadow Illusion," for instance, the background of squares *A* and *B* (*B* "belongs" to the lighter squares "on" the checkerboard, but is also "overshadowed" by a cylinder) means that we perceive *B* to be lighter than *A*. Once the background features are bracketed out by a uniform gray box, however, *A* and *B* present themselves as the same shade. The self-sufficiency of sensations is further undermined through Merleau-Ponty's observation that "synaesthetic perception is the rule" (*PP*, 229); carpets lack the same "redness," Merleau-Ponty argues, if found not to be woolly to the touch. Further examples are commonplace in our experience and the language we use to describe perceptions *more accurately*: the wet green of a spring lawn or the cold gray of steel, for instance.

Merleau-Ponty's point here is that empirical appeals to sensation misunderstand how perceptual objects are constituted. If sensations were "fully developed and determinate," at any given moment one should expect to experience a "*field of vision*"; a "segment of the world precisely delimited, surrounded by a zone of blackness, packed full of qualities with no interval between them, held together by definite relationships of size similar to those lying on the retina" (*PP*, 4–5). One doesn't, Merleau-Ponty contends, because experience is of a unified "perceptual domain" in which individual objects are charged with irreducible *meanings* codetermined by their relationships within a whole "phenomenal field" (*PP*, 47, 54). Contextualizing perception in this manner allows us to explain the phenomena just mentioned because, in this holistic context, each (apparently discrete) element "arouses the expectation of more than it contains" (*PP*, 4). One *expects* square *B* in the Checkershadow Illusion to be "darkened" by the cylinder's shadow, but not to the extent that it becomes one of the "dark" squares. Likewise, it is essential to the carpet's specific redness that one *expects* its woolly texture. Even objects lying outside one's "field of vision" retain existence within perception, Merleau-Ponty argues, as things one *expects* to see should one turn around (*PP*, 6). If these expectations were frustrated, the meanings constitutive of one's perceptual field would change. One's world would—literally—*be* different. This is exactly what happens once one brackets squares in the Checkershadow Illusion.

It will be useful, as far as the phenomena allow, to disambiguate which factors contribute to the respective diacritical backgrounds of "objects" perceived. Most straightforwardly, Merleau-Ponty is suggesting that empirical appeals to sensation misunderstand the interplay between features *of* our experience (e.g., interrelationships of color and/or texture) in allowing perceptual objects to stand out as they do. But there is another, more significant, contributory feature to their backgrounds: *our* active contribution, as a situated body-subject, *to* their characteristic features. I expand upon this important point in section 1.3, but Merleau-Ponty's basic contention is that the meaningful objects grasped in the preceding examples are *essentially* related to the abilities, projects, and commitments of the body-subject(s) in whose perceptual horizon(s) they show up. Merleau-Ponty's carpet's phenomenal redness, for instance, is partly determined by the expectations generated by *his* irreducibly *intentional* relation with it as "something on which to stand." This explains why its meanings (here: its color) would change should his expectations be frustrated by its coarse texture. Equally, the objects behind one's back exist within one's perceptual field as things (books, perhaps) that one might *intend* to soon appropriate. Finally, the squares in the Checker-shadow Illusion change shade depending on the expectations of someone hoping to *navigate* the scene before them. Since our expectations codetermine the intentional character of worldly entities, this explains why, as the Gestalt psychologist Kurt Koffka observes, as a matter of phenomenal priority, things often look "attractive or repulsive before [they look] black or blue, circular or square" (Koffka 1925, 320; *PP*, 24).

Merleau-Ponty is suggesting, then, that *meaning*, and not some more determinate constituent, is phenomenologically basic. This observation clashes with our default assumption that phenomena retain determinate qualities regardless of the fields in which they participate. Nevertheless, since this sort of "objective thought" is incompatible with the phenomena upon which it relies, when addressing the entities revealed in perception, "we must recognize the indeterminate as a positive phenomenon" (*PP*, 6).

To clarify this stronger claim about indeterminacy, we should explore where a shift in expectation corresponds to a more extreme shift in the figure-background relationship. Merleau-Ponty's paradigm example is of seeing a shipwreck. Since one doesn't expect to see mast poles protruding from the dunes—one might be out for a stroll, not hoping to appropriate a vessel—despite perceiving the relevant scene, one fails to see its elements *as* a ship. The immediate perceptual content is indeterminate and its elements only subsequently "fuse" to form a concrete structure (i.e., function *as* elements) as one approaches and a feeling of "uneasiness" (*PP*, 17) is resolved. Afterward, one cannot help but see the poles as belonging to that concrete structure. So, how could one miss these obvious clues as to the unity of the object? The answer is simple, Merleau-Ponty thinks, once we suspend objectivistic prejudice. The reasons for association between elements didn't precede the perception of the whole object as unified within the context of the overall field: "our perception ends in objects, and the object, once constituted, appears as the reason for all the experiences of it which we have had or could have" (*PP*, 67).

Negative support for this claim comes through Merleau-Ponty's observation that rival empirical explanations are unfeasibly fortuitous or circular. The reasons for association between perception's allegedly atomic constituents must remain absent, Merleau-Ponty explains, because distinct sensations cannot be thought of "as *the same*" without sacrificing their essential qualitative "this-ness" (*PP*, 15). Appealing to "extrinsic connections" with other sensations is also circular since "*de facto* proximities or likenesses" cannot plausibly *cause* perceptual atoms "to associate; it is on the contrary, because we perceive a grouping as a thing that the analytical attitude can then discern likenesses or proximities" (*PP*, 14, 16). Thus, the shipwreck experience previously outlined cannot be understood by appealing to "the same sensations differently associated" (*PP*, 16); it must involve a more fundamental shift in meaning generated by a shift in expectations. Merleau-Ponty puts this thought as follows: "When we come back to phenomena we find, as a basic layer of experience, a whole already pregnant with an irreducible meaning: not sensations with gaps between them . . . but the features, the layout of a landscape or a word, in spontaneous accord with the intentions of the moment, as with earlier experience" (*PP*, 21–22).

Merleau-Ponty uses the evidence cited to vouch in favor of Gestalt theory, and not some reductive empirical alternative. According to the Gestaltists, the furthest possible reduction in perceptual analysis is to the meaningful interplay of figure and background, rather than to discrete, context-independent sensations. Merleau-Ponty emphasizes that this irreducibility is not a contingent truth about factual perception, but "the very definition of the phenomenon of perception, that without which a phenomenon cannot be said to be perception at all" (*PP*, 4). Sensations simply couldn't collectively "form" our perceptual objects since seeing *anything at all* requires perceiving meanings bound up within a specific field. I expand on Merleau-Ponty's conception of the Gestalt in section 1.3. For now, it should suffice to say that, given that empirical sensations are incompatible with perception itself, we should jettison them from our account, since "the structure of actual perception alone can teach us what perception is" (*PP*, 4).

1.1.2 Objectivistic Prejudice

Merleau-Ponty thinks that empiricism makes what the Gestaltists call an "experience error" (*PP*, 5) by *reflectively* transposing the qualities of things perceived into perceptual consciousness. But this error isn't restricted to card-carrying empiricists. In an everyday sense, we know that we experience red carpets and telephone boxes, for instance, and erroneously believe ourselves to be directly acquainted with a determinate "redness" in our experiences of them. A related error also appears to be behind the usual binary yes/no question about whether the squares in the Checkershadow Illusion "really are" the same shade. By beginning from the presumption that *A* and *B* are determinate things, one passes on this prejudice to one's perceptions. Thus, we typically think that the primary constituents of perception themselves must retain (or just *be*) field-independent qualities.

However, as Kuhn helps emphasize, since perception provides our only means of access to the more-than-human world, the experience error has ramifications beyond perceptual analysis, because the insular dialectic it legitimates licenses an erroneous belief in the corresponding *ontological* determinacy of the things deemed responsible for our perceptions. The substantive problem here is that "we make perception out of things perceived. And since perceived things themselves are obviously accessible only through perception, we end up by understanding neither" (*PP*, 5). Since perception *ends* in objects, Merleau-Ponty argues, we may discern the world's physiognomy only by returning, as far as possible, to the perceptual phenomena themselves.

Taken this far, Merleau-Ponty's critique amounts to suggesting that a tacit adherence to empiricism's objectivistic presuppositions leads us to misconceive the targets of our concern, because we mistake perception's resultant "objects" for its fundamental constituents. More problematically, these objectivistic presuppositions license neglect of the diacritical background essential for any such objects to present themselves as they do. But Merleau-Ponty contends that the distorting influence of objective thought spreads further than this since, largely because of the experience error, the discrete objects finally registered in perception come to occupy a *privileged* ontological position. Objects, reflectively purified of their diacritical relationships, become the fundamental ontological constituents of the more-than-human world (*PP*, 71). Once conceived as such, Merleau-Ponty thinks, we become prone to overlook our objectifying tendencies and the reflective character of any objects we uncover by granting them the status of objects-in-*themselves*. That is, in conceiving of entities *as* objects we take ourselves to be acting in accordance with the basic preconditions of addressing them on their own terms. We even take the incompatibility of any such determinate objects with the phenomenological data (e.g., via the Gestaltists' experiments or the Checkershadow Illusion) to be a mark of *perception's* failure to attend to the facts: a tendency on the part of the epistemic "subject" to mis*represent* how the world "objectively" is. Attending to this shortcoming then becomes the privileged role of the natural sciences.

We see this sort of objectivistic prejudice manifested in naturalized accounts of the physiology of perception, which allegedly provide a "higher court of appeal" (*PP*, 7) than psychological analyses because physiology uses the "impersonal" and rigorously empirical third-person data of biologists, chemists, or physicists to establish the world's "objective" contents. According to the still-dominant physiological account (see chapter 3), bodily structures form an "anatomical path leading from a *receiver* through a definite *transmitter* to a recording station," where messages are decoded to "reproduce in us the original text" (*PP*, 7). This account licenses the well-known "reflex arc theory" (*PP*, 7) of perception and motor behavior as fundamentally mechanistic.[3] In principle, it also thereby supports the empirical "constancy hypothesis," which demands "a point-by-point correspondence and constant connection between the stimulus and the elementary perception" (*PP*, 7).

However, we've already seen reason to believe that the constancy hypothesis is irreconcilable with the phenomenological data. One cannot straightforwardly identify "the sensible" with the direct effects of external stimuli because, as Merleau-Ponty's carpet and shipwreck demonstrate, it is "bound up with a whole perceptual content" (*PP*, 8). By beginning from a prejudice toward the primacy of an "objective" world, however, naturalized physiological accounts repeat the objectivistic error that haunts the psychological account of sensation. Once one includes one's body in this picture—and why wouldn't one, if objects are fundamentally determinate and bodies (or their constituents) are themselves objects?—it is a short step to the physiological transmission model and the constancy hypothesis it legitimates.

The issue introduced *here* isn't only that, as a manifestation of objective thought, empiricism licenses reification of objects or properties-in-themselves as ontologically primitive, nor that it conceals the reflective character of this reification, but also that—by naturalizing the empiricist's previous acts of reification (i.e., by limiting explananda to objects-in-themselves)—the empiricist apparently requires us to understand the body as one more determinate object among others. One's body is, therefore, doomed to be subsumed under the restrictive, positivistic, terms that natural scientists afford to other worldly objects. We saw Charles S. Brown note in the introductory chapter that, according to the stipulations of our dominant scientific paradigm, doing so limits bodies to "*partes extra partes*" (*PP*, 73): a series of determinate extensional properties related only through a causal matrix.

But what exactly is wrong with this austere account? For one thing, the transmission account corresponds poorly with even the empirical data. Although there is observable "collaboration" (*PP*, 8) between the physiological sensory and motor systems, perceptual content isn't obviously affected by them in the requisite manner to license the theoretical posits above. Merleau-Ponty's famous analysis of Kurt Goldstein's patient, Schneider, demonstrates this. The transmission model must hold that Schneider's difficulties are the "mechanical" result of an injury to the occipital region, from which we should expect disruption to Schneider's visual field. But Schneider reports manifold other changes, including the inability to recognize familiar houses unless he intends to visit them (*PP*, 134). "Schneider's case," Merleau-Ponty argues, "shows deficiencies affecting the junction of sensitivity and significance, deficiencies which disclose the existential conditioning of both" (*PP*, 131).

Comparisons might be drawn with Merleau-Ponty's shipwreck. In both cases perceivers have sufficient sensory experience to discern the figure's "objective" structure ("ship" or "friend's house"), but don't (yet) do so because the available meanings are bound up with others within a perceptual field. Schneider's case is complicated by his injury, but this merely means that he loses the ability to recognize familiar places under certain circumstances. As Merleau-Ponty notes, Schneider *can* register the house if he intends to travel there because—as will become clearer in sections 1.3–1.4—a crucial factor in discerning meaning within the Gestalt is granted by the differential expectations generated by one's *motor intentional* relationship with the world. Such relationships, however, are incompatible with empiricism's underlying objectivistic stipulations, and these stipulations are imparted to any natural science it licenses. Thus, any attempt to return to "the things themselves" proves incompatible with empiricism because empiricism—as a manifestation of objective thought—"builds up all knowledge out of determinate qualities," offering "us objects purged of all ambiguity, pure and absolute, the ideal rather than the real themes of knowledge: in short it is compatible only with the lately developed superstructure of consciousness" (*PP*, 11).

1.1.3 Science Lacks Ontological Neutrality

As should be obvious, this "lately developed superstructure" roughly approximates the subject/object schema which, as Bruno Latour suggests, typifies modernity and continues to underpin natural science. It continues to do so because science retains a largely positivistic remit which seeks to exhaustively uncover the empirically real *objects* or properties allegedly manifested to a transcendent or transcendental *subject* (if reference is made to an observer at all).[4] This remains a common commitment amongst scientists even if, in a post-Kantian context, those objects may not correspond absolutely with the "noumenal" *things*-in-themselves. Natural scientists—at least insofar as they consistently claim epistemic access to the more-than-human world[5]—still aim to establish at least transcendentally purified access to the modality of the in-itself and the allegedly determinate "objects" (including any bodies) therein. One major problem with doing so, as we've seen, is that the "objects" thereby uncovered ordinarily misconstrue the phenomena presented to "the ultimate court of appeal in our knowledge of these things, our experience of them" (*PP*, 23).

Furthermore, to sustain the requisite universality of its epistemological contentions, even post-Kantian natural science still typically limits the contribution of the transcendental subject to her imposition of a limited range of *common* human concepts. Thus, the entities in which natural scientists trade retain no obvious reference to the specific diacritical features of the *concrete* phenomenological fields in which they are discerned. Natural science thereby gives the impression of addressing the objective "shape of the world" from the desituated viewpoint that Merleau-Ponty calls "high-altitude thought" and Donna J. Haraway calls "the God trick" (*PP*, 23; *VI*, 69; Haraway 1991, 189). From this transcendent or transcendental perspective, Merleau-Ponty thinks, the allegedly determinate ontological fragments that scientists *reflectively* discern—the physicist's quarks, the biologist's atoms, or the ecologist's energy flows—"will always appear more real than the historical and qualitative face of the world" (*PP*, 23) given in prereflective experience. This helps further explain why natural scientific terms achieve deliberative priority over the more partisan, transient, or nebulous (e.g., affective, aesthetic, or axiological) terms of phenomenological experience: the former allegedly score better in terms of "objectivity" because of their so-called perceiver-independence or neutrality.

One further problem with natural science's privileged status is that, as previously noted, by focusing squarely on the modality of the in-itself, science also limits the properties it can find in the world. As Brown (2003, 8) notes, within the naturalized discourse of extensional realism motivated by natural science's investigation of objects-in-themselves, "objectivity" conforms to a tripartite set of conditions requiring the scientist's target to be mind-independent (at least within the usual Kantian confines), wholly determinable, and admitting of only a single correct description. "Meaning" violates each of these criteria because it isn't obviously discernible independent of the Gestalt structure(s) in which it is expressed. Meaning is, therefore, typically exorcised from natural scientific accounts of more-than-human reality because it lacks objectivity, not because meaning doesn't show up as a basic constituent of experience, but because it apparently doesn't show up *in the right way*. Thus, Merleau-Ponty claims, science's positivistic and objectivistic lens limits or distorts the phenomena which, by right, should underwrite our ontologies and epistemologies. Nevertheless, because of our implicit commitment to objective thought, science enjoys higher authority on that very basis. Undercutting these objectivistic prejudices necessitates an attempt to return to the phenomena themselves, "and it is this pre-objective realm we have to explore in ourselves if we wish to understand sense experience" (*PP*, 12).

So, where are we? It should be clear that Merleau-Ponty's real target isn't any particular empirical account, but the objective thought which underpins it. Nevertheless if, as Merleau-Ponty and Kuhn claim, experience is the ultimate authority in even scientific attempts to understand the more-than-human world, then it *is* important to address the theories of perception implicit in any ultimate justification of the terms of debate. Simply put: if the subject/object schema—and, particularly, the presupposition that more-than-human reality is furnished by mind-independent objects-in-themselves—lacks phenomenological justification, as Merleau-Ponty suggests, then we have grounds for alarm, given that these objectivistic presuppositions underpin scientific naturalism. Moreover, if, as Merleau-Ponty suggests, natural science "is the second-order expression" (*PP*, viii) of phenomena, where scientific claims contradict the prescientific phenomenology upon which they rely (e.g., through the physiological transmission model, or denying meaning's ontological reality), we have reason to reject them. Since natural science lacks ontological neutrality, then, a genuine "return to the things themselves . . . is from the start a foreswearing of science" (*PP*, viii).

1.1.4 The Environmental Risks of an Unreflective Science

I've said little about the environmental crisis in this chapter. I've explored how a misplaced adherence to objective thought may underpin the objectivistic metaphysical stipulations inherited by natural scientists, but this alone says nothing about the violence I hold these stipulations to license. Before charting the route Merleau-Ponty prescribes to return to the more-than-human things themselves, then, it will be instructive to explore how the aforementioned stipulations might influence the attitudes and behaviors characteristic of our crisis situation.

We've seen Merleau-Ponty argue that objective thought leads natural scientists to reflectively reconfigure the phenomena upon which they rely. The larger problem with this is that the terms of scientific reflection render science substantively unable to question its own project, commitments, or boundaries. Merleau-Ponty puts this concern as follows: "Science manipulates things and gives up living in them. It makes its own limited models of things; operating upon these indices or variables to effect whatever transformations are permitted by their definition, it comes face to face with the real world only at rare intervals. Science is and always has been that admirably active, ingenious, and bold way of thinking whose fundamental bias is to treat everything as though it were an object-in-general—as though it meant nothing to us and yet was predestined for our use" (*EM*, 159). By failing to acknowledge its background commitments and presuppositions, science also presents everything as if it were "predestined for our use." The richness, indeterminacy, and alterity of the world is sacrificed from the outset by the reflective boundaries of scientific intelligibility through which something's "objective" nature may be appropriated by the suitably attentive observer.

Heidegger's analysis of "technology" helps highlight some dangers in this mode of revelation. Heidegger argues that technology, like natural science, is a particular "way of revealing" (1977, 5) the world that is motivated by the modernist schema under which epistemic subjects investigate an ideally separable ontic realm.[6] According to Heidegger: "The revealing that rules in modern technology is a challenging [*Herausfordern*], which puts to nature the unreasonable demand that it supply energy that can be extracted and stored as such" (1977, 6). We're familiar with its underpinning essence, or "Enframing" (1977, 9), via intensive agriculture where "soil" becomes shorthand for "mineral deposit," or industrial oil "fracking," where proposals turn on how much energy can be "unlocked." Other examples permeate our everyday interactions: think, for example, about commonplace references to "manpower" or "human resources." Through technological Enframing, *everything* is fundamentally characterized as "standing reserve" (1977, 8).

Technology limits the scope of ontology such that things can only really show up as wholly calculable collections of forces or energy reserves. Heidegger holds that technology, therefore, effectively reduces more-than-human entities to their use-value-for-us. But the error involved in technological Enframing is more-than-cognitive. By setting the metaphysical boundaries it does, technology dictates the shape that affective or behavioral engagement with those entities takes. If rainforests, for example, are reducible to the goods (e.g., timber, fuel, bushmeat) and services they may provide for us (e.g., by regulating local climates or "recycling" carbon dioxide emissions), then, since they *are* resources, we will inevitably relate to them as primarily things to be used.

What Heidegger helps us tease out is that, under technology at least, the behavioral violence emblematic of our crisis situation (aggressive deforestation, excessive fuel consumption, and so on) appears to be nonaccidentally related to the violence already implicit in the terms of debate. Bryan E. Bannon (2014, 64) notes that technological Enframing effectively naturalizes "usefulness" as an ontologically real property-in-itself. In fact, "usefulness" (variously conceived) becomes *the* definitive property of objects-in-general. Bannon argues that technology thereby conceals the asymmetrical relationship between "resource" and "beneficiary" behind any such use-value and, ultimately, naturalizes *mastery* as one's ordinary means of relating to the more-than-human world. In this context, there can be no clear distinction between violent actions (practice) and the conceptual means (theory) by which those violent acts are licensed. This is because theory delineates the ways we relate to the entities it posits as a matter of praxis. But it is also because technology's utilitarian metaphysics arises, to some degree, from the partial and interested ways that we *already* habitually relate to the rest of the world (i.e., as objects to appropriate).

How is this salient to our present debate? Even conceived technologically, sustained engagement with "environmental issues" isn't impossible; it is just limited to "solutions" regarding effective *resource* allocation or sustainable *consumption,* because nonhuman "objects" are ontologically reducible *to* "natural resources." According to Heidegger's understanding, the technoscientific environmentalist can be an environmentalist only insofar as she concerns herself with how to meet fuel, water, or protein needs more sustainably. She simply isn't free, however, to relate to the rainforest other than as a storehouse or "glorified septic tank" (Evernden 2003, 22) because, for her, the rainforest *just is* its utility. Thus, if some of the problematic behaviors already mentioned can be traced back to this sort of hierarchical, utilitarian purview that Neil Evernden calls "resourcism" (2003, 22), then technoscientific "solutions" like those we've seen Paul Crutzen prescribe—*consuming* less meat, *utilizing* bioadaptive technologies, *managing* natural *resources*—can play only a limited role in disrupting our environmental crisis. If theory and praxis are as intimately connected as they appear to be, then it is implausible to expect a colonial conceptual scheme to bring about the anticolonial behaviors and attitudes required to mitigate our crisis situation at root. To cite Audre Lorde's dictum: "the master's tools will never dismantle the master's house" (1983, 94).

One might nonetheless wonder how this discussion of technology relates to our critique of naturalism, given that technology and natural science aren't straightforwardly synonymous. The committed ecologist or biologist, for example, would rightly baulk at the suggestion that she must identify all ontologically real entities (marine ecosystem, sea cucumber, eel's digestive tract, and so on) with their use-value *for human beings*. However, Heidegger emphasizes that technological Enframing relies upon the reductive descriptions that science affords: "Modern science's way of representing pursues and entraps nature as a calculable coherence of forces. Modern physics is not experimental physics because it applies apparatus to the questioning of nature. Rather the reverse is true. Because physics, indeed already as pure theory, sets nature up to exhibit itself as a coherence of forces calculable in advance, it therefore orders its experiments precisely for the purpose of asking whether and how nature reports itself when set up in this way" (1977, 10).

Heidegger is here driving at a point similar to Merleau-Ponty's. He is suggesting that natural scientists cannot begin from a neutral starting point in their attempts to derive knowledge about the more-than-human world because doing so requires a *determinate* field of enquiry and, therefore, the inauguration of background conditions which determine that field. These conditions subsequently dictate the appropriate epistemological apparatuses and truth conditions, and, ultimately, determine how science's target objects show up for observation. Like Merleau-Ponty, Heidegger argues that natural science cannot, therefore, straightforwardly arbitrate the distinction of real from unreal entities—and, thus, provide the privileged ontological grounds for environmental philosophy—because science doesn't dispassionately address the world independently of the background commitments which determine its findings. Indeed, Heidegger concurs with Merleau-Ponty that, by reifying things-in-themselves as the basic element available for analysis, the scientist already reflectively distorts the underlying phenomena and admits only problematically restrictive means of epistemic access to them.

If, as Merleau-Ponty (*PP*, viii) and Heidegger (1999, 102) both claim, natural science is motivated by the desire to achieve a systematic and *totalizing* account of the more-than-human world, then this would explain why science (and not just technology) limits inquiry to a wholly determinable realm of things-in-themselves. Given that contemporary science effectively defines these entities in terms of their *causal* properties, however, then the connections between science and technology are more intimate, and potentially dangerous, than they first appear to be. Not only are meaning, indeterminacy, (nonhuman) agency, and alterity made antagonistic to science's basic explanatory project, but nonhuman entities are permitted to exist in virtue of a remarkably narrow set of extensional properties, which are nonaccidentally related to the method or task set for them—by the scientist—from the outset.

Differently put: if, as Merleau-Ponty suggests, natural science is motivated by a particular explanatory project (i.e., to more effectively navigate or better utilize worldly "objects" as part of an intersubjective community), then science's findings cannot be entirely detached from the utilitarian assumptions and apparatuses partly constitutive of that background. Nor can they be entirely detached from the objectivistic assumption of a realm of wholly determinate and determinable objects-in-themselves; an assumption which derives largely from a misunderstanding of perception, yet which provides the background legitimacy of natural science's totalizing project. More problematically, these assumptions will also delineate any "solutions" licensed by understanding environmental "issues" as science dictates.

Although these concerns might sound curiously abstract, it isn't difficult to find examples to substantiate them. I want to focus here on a less obvious one: where ecological terms are appropriated by the ecosystem services literature. At first glance, this literature promises more positive and self-reflexive ways to understand and relate to the more-than-human world. While the (e.g., aesthetic, nutritional, or geochemical) services that ecosystemic entities provide within such frameworks are almost always made with reference to a human beneficiary, these asymmetric use-relationships aren't surreptitiously enframed as properties of things-in-themselves. Rather, the asymmetry between entities is emphasized as part of what constitutes a valuable "service" in the first place (Jax et al. 2013, 261). Furthermore, in principle, ecosystem services approaches allow more-than-human entities to be valued noninstrumentally (as in the case of nonsubstitutable sacred landscapes), and even recognize "meaning" (e.g., a glacier's aesthetic value) to be metaphysically real (Jax et al. 2013, 262).

Nevertheless, ecosystem services approaches remain problematic because their basic ontological presuppositions, if not their whole ontology, are inherited from the ecological science employed to motivate biodiversity preservation in conservation biology (Jax et al. 2013, 264). Within this purview, species biodiversity, for instance, is valuable only insofar as its reduction has the potential for a demonstrably negative (i.e., causal and measurable) impact on the wellbeing of the ecosystem or subsystem. Ecosystem services approaches simply seek to extend the kinds of wellbeing under consideration. What remains, therefore, *is* a reduction of more-than-human entities to natural *resources*—in the sense of being reducible to their causal powers—if not straightforwardly to resources-*for-us*. But the problem here isn't just that the ecosystem services concept was born of a project investigating more sophisticated economistic analyses of conservation initiatives. A deeper issue persists because, as Bannon (2014, 68) argues, the totalizing assumption that more-than-human reality consists of a set of wholly calculable or orderable forces is the grounding principle upon which ecological science rests and so simply isn't up for revision in *any* extension or application of it.

What is interesting about the ecosystem services literature, then, is that it can restore concepts like meaning and value to the more-than-human world, but only at second remove. Any such concepts are, therefore, colored by the implicitly utilitarian background against which things emerge as ontological entities for the ecologist in the first place. To reiterate: this is because, as Heidegger and Merleau-Ponty both argue, the ecologist's objectivistic metaphysical presuppositions are intimately related to the projects and methods under which "the natural world" is inaugurated as a determinate field of inquiry. The background against which ecological science becomes intelligible seemingly requires the epistemic investigator to conceptualize and, ultimately, relate to all more-than-human entities *as* collections of latent causal powers (i.e., resources). It isn't entirely surprising, therefore, that the ecosystem services approach readily lends itself to straightforward monetization and other economistic and technoscientific means of commodifying the more-than-human world (Jax et al. 2013, 261).

Of course, this doesn't mean that ecologists are necessarily *bound* to enframe the world in Heidegger's more obviously problematic anthropocentric sense. Heidegger concedes that there is no straightforward entailment relationship between natural science and technological Enframing. Nevertheless, the dominant illusion that the latter is the logical consequence of the former "can maintain itself only so long as neither the essential origin of modern science nor indeed the essence of modern technology is adequately found out through questioning" (Heidegger 1977, 11). This questioning must take place at a metatheoretical level and cannot, therefore, be carried out by science itself. Given its narrow sphere of concern, "science does not think" (Heidegger

1977, 8). An interrogation of the violence implicit in our "submerged mass" must aim to situate itself outside scientific naturalism's objectivistic and utilitarian prejudices. Genuinely radical reflection must be—at least *from the start*—a foreswearing of science.

A word of caution might be wise. Although Heidegger is sometimes interpreted in this way, it would be disastrous for environmental philosophers to adopt an antiscientific (rather than anti*scientistic*) viewpoint. As Alison Stone notes: "Ecophenomenologists should not throw modern science overboard. We owe the sciences our knowledge of climate change, species and habitat loss, ocean and soil degradation, and other facets of the ecological crisis. Having said that . . . we also need to criticise the technoscientific project of knowing and controlling the whole of nature" (2015, 2). In this spirit, like Merleau-Ponty, I suggest that we take natural science to provide only *regional* understanding of a world already carved up according to its stipulated background terms and methods of enquiry. Doing so allows science to remain important in understanding the contours of our environmental crisis. However, doing so also requires ceding the *basic* court of appeal in ontological and epistemological matters to the phenomenological contexts in which scientific insights are always garnered. In this sense, natural science might be said to return to "the 'there is' which underlies it; the site, the soil of the sensible and opened world such as it is in our life and for our body—not that possible body which we may legitimately think of as an information machine but that actual body I call mine" (*EM*, 160).

"In this primal historicity," Merleau-Ponty thinks, "science's agile and improvisatory thought will learn to ground itself upon things themselves and upon itself, and will once more become philosophy" (*EM*, 161). We shall soon see how ecophenomenology might marry scientific insights with a more robust method of (self-)reflection. It is enough, for now, to highlight the dangers inherent in natural science's underlying objectivistic commitments without disregarding its contributions to knowledge.

1.1.5 Why Intellectualism Won't Help

I have suggested that natural science distorts the phenomena which open its possibility, and thereby becomes prone to legitimate a problematically utilitarian worldview that science is itself ill equipped to address. Primary responsibility for these concerns appears to lie in its adherence to objective thought, which licenses the violence manifested in a utilitarian or objectivistic praxis by neglecting the perceiver's contribution to the basic physiognomy of objects encountered. Perhaps, then, science's problem lies in its empiricism. Under empiricism, "attention" is "a general and unconditioned power . . . applied indifferently to any content of consciousness" (*PP*, 26), but this proves insufficient to explain the intertwining relationships involved in the determination of perceptual objects. By failing to acknowledge the body-subject's intentional contribution to natural science's background context, empiricism distorts the phenomena upon which it relies and reifies its own way of revealing as universal or "acosmic" (*PP*, 24). As we have seen, empiricism may also thereby serve to perpetuate anthropogenic violence by naturalizing hierarchical and utilitarian ways of relating to the more-than-human world.

By making attention an active faculty, however, "intellectualism" seeks to acknowledge the contribution of the perceiver required for the entities described by natural science to present themselves with the physiognomy they do (*PP*, 27). "Intellectualism" refers to the Kantian or Cartesian rationalism under which "perception must be organized by, indeed it just *is*, thought or judgment" (Carman 2005, 59). In Descartes's case, for instance, appropriately focused attention illuminates the wax's essential qualities as described by Galilean physics. One also perceives, but doesn't straightforwardly see, the men outside hidden by hats and coats. One *judges* them to be there (1996, 20–21). Likewise, for Kant, one's perception of a cube is governed by a priori conceptual structures which offer lawlike rules for how an object must appear to a situated subject given their relation in "objective space" (*PP*, 300–301; Jensen 2013, 50). Intellectualism, therefore, retains a key focus on the contribution of perceiver to objects perceived, which is denied under empiricism to its detriment. Furthermore, since it is compatible with naturalism, a successful intellectualism might legitimate (intellectualist) natural science's continued stronghold on the "material" or "contentful" aspects of our knowledge of the more-than-human world.

The main problem with intellectualism, however, is that "since in attention I experience an elucidation *of* the object, the perceived object must already contain the intelligible structure which it reveals" (*PP*, 27). Merleau-Ponty's point is that, if intellectualism may avoid a vicious regress under which any account of the origin of perceptual content is forever omitted, the intellectualist is left with two equally unsatisfactory options. The first is to concede that the world's intentional physiognomy is merely projected or constructed by the epistemic subject. Neither Kant or Descartes, nor their heirs in this tradition (including Meillassoux), will accept this option since it effectively denies the mind-independence of the entities science investigates. Given that our present concern is partly to do with whether natural science is equipped to address the more-than-human world on its *own* terms, this option should be ruled out a priori.

Intellectualists must opt for the other option: to "maintain an element of the empiricist idea of a manifold of sensory matter which is synthesized by the understanding" (Jensen 2013, 51). Even Kant, who rejects any ontologically real "sensations" in the empiricist mold, concedes that one must appeal to an "*ideally separable* sensuous moment of perception" (Jensen 2013, 51) to retain the requisite realism about the mind-independent world. However, Rasmus Thybo Jensen notes that the issue arises, for Merleau-Ponty, that if Kantian intuitions were blind (i.e., meaning*less*), then they couldn't guide the synthetic activities of the understanding. Judgment would become the "indifferent linking of objects," and the concepts introduced to explain how intuitions are guided or associated couldn't avoid being "empty" and, therefore, redundant in their task (*PP*, 32). Something more prescriptive is needed. For Kant, this comes through understanding Newton and Descartes's reductive, mechanistic principles to be legislative a prioris, somehow "*of* Nature" (*N*, 23), which determine how the more-than-human world must be revealed by the understanding.

The second option, therefore, suffers a predictable problem: by taking up an empiricist account of the sensory manifold, intellectualism appears bound to inherit empiricism's objectivistic assumptions. Since its account of perception derives from the "sensations" of empiricism, even Kantian intellectualism ultimately concedes the very objective thought which haunts empiricism, albeit in the mode of "empirical realism." Empiricism and intellectualism, therefore, "have this idea in common, that attention creates nothing, since a world of impressions in itself or a universe of determining thought are equally independent of the action of mind" (*PP*, 28). Post-Kantian science results in little more than resituating precritical scientific realism in the context of transcendental idealism and, therefore, represents very little interrogation of its objectivistic and utilitarian terms. Meaning continues to be ruled out as a basic feature of the world *itself*, not because attentive judgment rightly demands it, but because intellectualism rests on the same objectivistic misunderstandings about phenomena that empiricism does.

Intellectualism cannot solve science's problems because it conflates perception and judgment. Intellectualism also thereby seems to refute itself. Once one "realizes" that the squares in the Checkershadow Illusion, for instance, "are" the same color—that is, the "objective" stimuli are the same color—one judges them to be so. Nevertheless, one continues to experience their colors *as* indeterminate, fluctuating between shades. This suggests an *inherent* distinction between perception and judgment, because "things do not always *really* appear to me the way I *think* they appear, and intellectualism can make no sense of that distinction" (Carman 2005, 63). Thus, "to perceive in the full sense of the word," it seems, "is not to judge, it is to apprehend an immanent sense in the sensible before judgement begins" (*PP*, 35). What intellectualism misses no less than empiricism is that the meanings which we receive *in* perception, and which are essential to *guide* judgment, must constitute a constituent component *of* perceptions themselves.

Intellectualism's problems ultimately derive from its account of the transcendental subject presiding freely over objects-in-themselves. "It thus does away with all problems except one: that of its own beginning" (*PP*, 38). By moving from the radical "objectivity" of empiricism to the radical "subjectivity" of intellectualism, the allegedly determinate exteriority of the "objective" world is left intact. But we lose even the (untenable) transmission account of how one comes to perceive people, sea cucumbers, or carbon deposits because "perception" becomes "thought about perception." A concerted attempt to return to the things themselves, then, requires an approach to perception which rejects these cleavages. One must account for the fundamental "subjectivity"—here: something like "concrete intentional meanings"—of perception abandoned by the former. One must also retain a plausible account of prereflective phenomena, which is lost in the latter. Given that both errors appear to be grounded in a basic commitment to objective thought and the subject/object schema it legitimates, our task requires problematizing the binarism bequeathed by modernism. Doing so becomes possible by acknowledging the intentional salience of one's *embodiment* in establishing the phenomena which underwrite our theoretical contentions. In what follows, I will explore how Merleau-Pontian ecophenomenology might juggle these requirements through its radicalization of Gestalt theory.

1.2 THE GESTALT

The issues surveyed above appear to be largely traceable to the metaphysical prejudices licensed by objective thought. Since scientific naturalism relies upon the objectivistic assumptions under scrutiny, an attempt to "return to the things themselves" must begin by suspending the presumption of naturalistic metaphysics. Gestalt theory, however, shows greater promise because it registers the primary importance of *form* or *structure* in establishing ontological or epistemological claims from phenomenological experience.

Before we turn to the specifics of Merleau-Pontian Gestalt theory, we should field a potential concern. The contemporary reader might find the appeal to classical Gestalt psychology a strange source of justification given its unpopularity among contemporary psychologists and cognitive scientists. It is noteworthy, however, that Gestalt psychology hasn't become unpopular because it contends that perceptual phenomena cannot be reduced beyond figure-background relations. It is criticized, rather, because of its apparent inability to explain those phenomena causally, and because its findings resist precise quantitative measurement. The former concern is one which Merleau-Ponty explicitly accepts. Gestalts cannot be explanatory in any simple causal sense because we simply cannot make sense of putative causal phenomena more mereologically basic than Gestalt-objects (*SB*, 143–44). The latter concern has recently been challenged. Moreover, if, as the literature suggests (e.g., Jäkel et al. 2016, 5; Wagemans et al. 2012, 1219), both concerns boil down to a failure to cohere with the reductive and mechanistic explanatory framework of contemporary cognitive neuroscience, then this "failure" needn't be fatal, especially given the role that its objectivistic terms may play in licensing the violence emblematic of our present environmental context.

Nonetheless, if perception provides our only means of access to the more-than-human world and, perceptually speaking, the Gestalt is irreducible, then we should resist the urge to situate the Gestalt in naturalistic terms. Merleau-Ponty accuses the classical Gestaltists of this error. He argues that classical Gestalt psychology retains a certain neo-Kantian intellectualism insofar as it seeks to establish ontologically determinate objects which are, paradoxically, inaccessible to perceiving subjects on their own (i.e., noncorrelational) terms (*SB*, 224). This criticism is important, because the shift in approach required to appreciate the Gestalt-situatedness of our ontological and epistemological contentions shapes the ecophenomenological recommendations that follow. Given the *environmental* focus of this book, however, it will be instructive to explore how this same error impedes the work of revolutionary natural scientists like Kuhn, who might otherwise be thought to evade the various concerns hitherto identified.

Kuhn explores Gestalt-shift in the context of scientific paradigm shift. For Kuhn, although scientists may perceive phenomena differently, "scientists do not see something *as* something else; instead they simply see it" (1970, 85).

Kuhn's point isn't that all scientists receive qualitatively identical sense data and merely judge it differently. That is the "usual view" (1970, 121) of science, which Kuhn chastises Descartes for inaugurating. Kuhn's point is that scientists differentially perceive *real* objects based on their (meta)theoretical presuppositions and experimental apparatuses. Nevertheless, for Kuhn, "the world does not change"; there exists "the same constellation of objects," only subsequently carved up according the intentional requirements of differing scientific paradigms (1970, 121–22). Kuhn acknowledges that Aristotle and Galileo registered differing phenomena depending on their conceptual manipulation as either "pendulums" or "constrained falls," for example. However, Kuhn argues, both still see *the* swinging stones—or their alleged constituents: the "many related chemical, electrical or dynamical phenomena" (1970, 129)—in the same way that one might see a duck or rabbit, yet the lines on the page exist determinately as things-in-themselves. Despite Kuhn's insistence that scientific claims rely upon their underlying perceptual phenomena, then, even naturalized Gestalt theories like his remain problematically beholden to the determinate objects that objective thought mistakes for the primary constituents of ontological reality.

In a familiar move, Merleau-Ponty emphasizes that "it is from the universe of perceived things that Gestalt theory borrows its notion of form" (*SB*, 144). Through their tacit intellectualism, Kuhn and the Gestaltists made a category mistake by failing to realize that "in the final analysis form cannot be defined in terms of reality but in terms of knowledge, not a thing of the physical world but as a perceived whole" (*SB*, 143). Given the irreducibility of the Gestalt to something more perceptually basic, and the concession that epistemology and ontology rely absolutely on perception, neither the Gestalt nor its allegedly determinate constituents can be unproblematically taken to exist *within* the world-in-itself (noumenal or otherwise). The Gestalt must, rather, exist for the perceiver as the unity of her perceptual wholes; perceptual consciousness must be a Gestalt like any other.

Merleau-Ponty's point is subtle, but essential to ecophenomenology's potential to address the objectivistic assumptions which pervade natural science and, ultimately, to disrupt the violence which fundamental commitment naturalism licenses. To illustrate, recall Jakob von Uexküll's ticks, for whom the world's meanings emerge in a manner intimately related to the kind of entity it *is*. Given its physiological and functional orientation, the "smell" of butyric acid exists *as* a phenomenon and the taste of raisins does not. That is not to say that the tick isn't interested in the raisins, but that "raisins" don't form part of the composite structure of the tick's meaningful phenomenal world in the Kantian sense Uexküll officially invokes. The problem we saw Uexküll identify in the introductory chapter becomes important here, because the tick's world is intractable from its immanent participation in its own *Umwelt* or (eschewing neo-Kantian connotations) Gestalt. For natural science to accurately and exhaustively describe the tick, its world, or *the* "objective" world from which the tick's derives, then, would require metaphysical speculation from the viewpoint of acosmic consciousness that Merleau-Ponty rightly problematizes. Likewise, for natural science to exhaustively determine the "mind-independent" reality of ticks, rainforests, or marine ecosystems requires the problematic assumption of an objective world with which one, as an epistemic subject, is somehow transcendentally acquainted. More basically, it requires misplaced adherence to the background stipulations of objective thought under which butyric acid, raisins, rainforests, and marine ecosystems are representations or constellations of ontologically determinate phenomena. Not only are each of these commitments misplaced, insofar as they implicitly license the colonial imposition of a naturalistic metaphysics, they also hamper our attempts to address the violent attitudes and behaviors responsible for our crisis situation.

Some mainstream environmental theorists have attempted to rehabilitate naturalism by taking up insights like Uexküll's from within its confines. But they only serve to illustrate the tensions between acknowledging our Gestalt-situatedness and the sort of naturalism which ordinarily delineates our ontological and epistemological contentions. The sociobiologist E. O. Wilson, for instance, formally subscribes to a radical neo-Kantianism under which "the idiosyncrasies of human evolution" dictate how the world is represented to consciousness via "sensory input and the self-assembly of concepts" in the brain (Wilson 1998, 66). Despite formally exorcizing any free-floating Kantian "subject," however, Wilson claims that "*the proper task of scientists is to diagnose and correct the misalignment*" between "outer existence [and] its inner representation" (1998, 66). As Ted Toadvine (2009, 11–12) also observes, Wilson thus attributes to natural science the privileged possibility of transcendent access to the mind-independent world which he otherwise denies to body-subjects—paradoxically, *including* scientists—because of their material implication in the world's causal structures. Given my warnings about the objectivistic and utilitarian commitments which pervade natural science, it is relatively unsurprising that Wilson ends up espousing a suspiciously univocal, mechanistic, and deterministic account of the world from his paradoxically "acosmic" standpoint.

What Merleau-Ponty's appropriation of Gestalt theory allows us to better appreciate, then, is that, by taking up naturalism's objectivistic heritage, one risks doing colonial violence to the more-than-human world by relatively uncritically extending the meanings available within a particular kind of Gestalt—meanings intimately related to the specific kinds of entities *we* are—to a world which might offer other meanings to differently configured entities. Moreover, if my previous contentions about the partisan commitments behind natural science's totalizing schema are on the right lines, then the meanings which show up for natural scientists—like Uexküll's ticks—will be nonaccidentally related to the asymmetric interests and projects which motivate their investigation of the more-than-human world in the first place.

Fully acknowledging the Gestalt-situatedness of our epistemological and ontological contentions appears to be incompatible with our default commitment to naturalism. Nonetheless, the question of how to uncover the *more-than*-human world on its own terms cannot be solved by denying access to it in the correlationist manner that a wholly consistent Kantianism might suggest. This sort of position is problematic in that it *still* begins from the subject/object schema under which more-than-human "objects-in-themselves" must be either readily accessible (as in dogmatic naturalism) or foreclosed (substituted for a realm of correlational objects-for-us). As I hope to have shown, part of the problem with the supposition of this objectivistic schema lies in its misunderstanding of the perceptual phenomena upon which it is based. Perception *ends,* rather than begins, with the determinate objects of its concern. The presumption of a primitive ontological realm of objects-in-themselves is, therefore, a misleading and unhelpful abstraction.

Merleau-Ponty thinks we should instead resituate our ontological and epistemological investigations within the context of the meaningful Gestalts which allow phenomena to stand out as they do. Merleau-Ponty intends to thereby disrupt the false dilemma between naturalism and correlationism which might otherwise preclude access to the more-than-human world on its own rich terms, and/or neglect the commitments and assumptions which contribute to the physiognomy of the "objects" revealed perceptually. This becomes plausible because, in the Gestalt context, meaningful phenomena are *coexpressed* ambiguously *between* world and body-subject—be they a tick or a human being—and, therefore, not wholly beyond our reach. By resisting a fundamental subject/object cleavage, theory is "no longer Kantian" (*N*, 169) in the problematic sense outlined above. Otherwise put: while acknowledging our Gestalt-situatedness problematizes granting a dogmatic or positivistic science a monopoly on ontological and epistemological matters (especially where we seek to reduce anthropogenic violence), doing so needn't cleave us from the more-than-human targets of our concern because of how phenomena are instituted within this context.

In what follows, then, I discuss Merleau-Ponty's notion of the Gestalt, and its implications for our attempt to better address the world on its own terms. By foregrounding the Gestalt-situatedness of our onto-epistemological contentions, I hope to make possible several avenues of progress in addressing the submerged mass of our crisis situation. The first is straightforward: by understanding all (scientific) worldly engagements within the Gestalt structure rather than any derivative subject/object schema, Merleau-Ponty licenses more positive, and less violent, ways to understand and relate to the more-than-human world. I contend that this benefit arises largely because these Gestalt*s* (i.e., plural) legitimate a richer variety of ontologically real meanings, and associated means by which to relate to them, than might otherwise be admitted via natural scientific apparatuses. Secondly, and perhaps more importantly, by contextualizing *our own investigations* within certain kinds of Gestalt structure, Merleau-Ponty forces a *radical* critical self-reflexivity about our imposition of onto-epistemological violence. Thus, I want to claim, disrupting the latent violence of our submerged mass becomes more plausible through the radically reflective mindset necessitated by taking seriously the Gestalt-situatedness of our reflective observations or theoretical contentions.

1.2.1 Embodiment and the Gestalt

Given that Gestalt theory is primarily a theory of perception, let's briefly recap the flaws with the objectivistic accounts that led us to it. We saw that empiricism effectively reduces perception to receipt of the stimuli transmitted by the mechanical structures of one's body. Intellectualism reduces perception to intellectual judgment yet, insofar as it is intelligible, intellectualism requires the body to carry out a similar causal role to account for the "sensations" it shares with empiricism. Neither manifestation of objective thought grants any primary importance to perceptual *embodiment*, a debt Merleau-Ponty seeks to pay by embedding the perceiver *within* the Gestalt.

Toadvine (2009) explains that Merleau-Ponty's concept of the Gestalt consists of three fundamental characteristics, none of which, independent of an antecedent commitment to the subject/object schema, requires limiting perceptual consciousness to humanoid body-subjects:[7]

1. There exists a dialectical or diacritical relationship between parts of the Gestalt and the whole.
2. There is a teleological element which constitutes normativity in the Gestalt.
3. There are degrees of complexity to which the Gestalt is made thematic and may be transposed.

We best see how these characteristics play out through what Merleau-Ponty calls "vital behavior": an organism's environmental interaction. The first two are especially difficult to separate. To illustrate, Merleau-Ponty draws upon the observation that, by amending its ambulatory rhythm, a dung beetle can walk almost immediately after having a leg amputated. By reorganizing several elements (individual leg movement, rhythm of combined leg movement, whole body movement, and so on), through varying dialectics between (parts of) the organism and its environment, the beetle can achieve its vital goal: to walk more effectively. There are, therefore, better and worse ways for the beetle to respond to *its* milieu, and the reorganization of various microlevel dialectical relationships can only really be understood in this teleological context.

Of course, any such teleology is here registered by situated human observers. Nonetheless, it cannot be entirely human projection. The beetle's goal is rather a "norm enscribed in the facts themselves" (*SB*, 123), essential to understanding its functional reorganization and behavioral orientation. After all, notwithstanding prior commitment to objective thought, the beetle needn't be mechanically responding to the mind-independent stimuli described by (an ideal) science, nor need the beetle's world be entirely "subjective" and, thus, foreclosed to us. In fact, neither of these explanations appears adequate to *make sense* of the observed phenomena. Like Uexküll's ticks, Merleau-Ponty's beetle appears to respond "to the meaning, the *sense*, presented to it by its situation, even though this meaning is not a conscious representation" (Toadvine 2009, 27). Phenomenologically speaking, this sort of meaning isn't problematic for us to appreciate, even though the beetle's goal, like the "smell" of butyric acid, cannot show up for us in the same way it does for the animal in question. Acknowledging the normativity intrinsic to the beetle's Gestalt is important, however, in allowing us to appreciate the "directed activity between blind mechanism and intelligent behaviour which is not accounted for by classical mechanism and intellectualism" (*SB*, 40).

The third characteristic relates to the establishment of what is typical of these dialectical relationships such that an organism can transpose its meaning elsewhere, something that becomes difficult for natural science to explain in lieu of any straightforward causal explanation. Merleau-Ponty draws upon the observation that a cat which feeds itself by tugging a string with its paw, on the second attempt, often uses its teeth with greater success. This behavior cannot be satisfactorily explained through the reflex arc because the cat's behavior, like the beetle's, constitutes a "new aptitude for resolving a series of problems of the same form" (*SB*, 96). It doesn't mechanically repeat gestures, given the repetition of qualitatively identical stimuli. We've seen reason to believe that the cat's behavior might, however, be understood via reference to the diacritical features and embodied expectations generative of its field. To borrow Merleau-Ponty's imagery, even for cats and beetles, the meaning of each note within a Gestalt is to be found only within the symphonic whole. Like any symphony, however, its "melodies" may be transposed into a different key.

Why is this important for our task? Merleau-Ponty's research suggests that the object/subject distinctions which underpin natural science and, subsequently, delineate our understanding of the environmental crisis aren't just reflective abstractions, they're also "badly made" (*SB*, 10). Natural science typically precludes meaning from its account of ontological reality because, apparently unlike the world's more "objective" furniture, meanings cannot be identified as positive things-in-themselves. However, the examples discussed problematize this exclusion. Meaning isn't so self-evidently subjective (or fictitious) since an appreciation of contextual meaning appears essential to grasp tick or dung beetle behaviors. But neither is meaning self-evidently reducible to, or emergent from, the allegedly determinate objects or properties natural science describes. As Merleau-Ponty's shipwreck and Uexküll's ticks neatly demonstrate, configuring these objects or properties *as* stimuli also requires the contribution of situated organisms within an overall Gestalt structure. Moreover, the world's more fundamental "objectivity" is thrown into question, given that understanding vital behavior requires not, as even Kuhn contests, an understanding of *the* determinate physico-chemical properties or objects (bodies included), but an appreciation of meanings inseparable from the vital milieu into which an organism is materially embedded. Finally, given the irreducible teleological normativity of Gestalts, perceptual acts and their resultant objects aren't "objective" in the sense of being epistemically neutral or value-free, even for nonhuman animals. If so, then, by investigating a reified realm of objects-in-themselves, natural science violently excludes a richer range of meanings from its investigations than is legitimated by the perceptual phenomena themselves. If axiological value, for instance, cannot be exorcised from insect, feline, or arachnid worlds without significant distortion, then there is no obvious reason why such meanings should be excluded from our broader investigation of the more-than-human world, nor from our investigation of the entities and relationships which constitute our environmental crisis.

But the biggest ramifications concern how we understand *our* bodies' roles in configuring the meaningful entities which furnish our worlds, and which constitute explananda in our reflective accounts of them. Merleau-Ponty takes the examples above to show that, while mechanistic reflexes may be "a very special case of behaviour, observable under certain determined conditions" (*SB*, 46), one cannot extrapolate totalizing explanations from these exceptional events. In the Gestalt context, "fundamental" physical laws can "have meaning only as a means of conceptualizing the perceived world"; they exist as "certain privileged perceptual structures," rather than providing *the* privileged transcendental structures systematically constraining ontological pronouncements (*SB*, 144–45). This observation is problematic for contemporary natural science, given that it seeks to provide exhaustive explanations of worldly events through determinate extensional properties linked via causal matrices.[8] Merleau-Ponty's examples suggest, however, that the specific composition of one's body and its modes of comportment into the world are intimately related to the more fundamental determination of any such phenomena. One cannot, therefore, unproblematically impose these explanatory structures from outside. The tick's "objectification" of mammalian blood is irreducibly related to its material and functional organization, for instance, as is the beetle's capacity to reorganize its limbs for better ambulatory movement. Similar conclusions are commonplace in the (bio)enactivism literature.[9] "Even at the minimal cellular level," Darian Meacham writes, the "chemically necessary dynamics of the environment do not fully determine the dynamics of the internal milieu or the behaviour of the system" (2016, 80). Meacham suggests, therefore, that even protein behaviors should be considered "cognitive" in some sense that is nevertheless inseparable from the kind of material entity that the protein *is*.

But these revelations also give *us* a problem. They force us to recognize that, for even relatively simple organisms, their basic *bodily* relationship with the rest of the world isn't straightforwardly causal, but irreducibly *intentional*.

Understood as such, "subject" and "object" are "two 'abstract' moments of a unique structure" (*PP*, 430), only subsequently individuated by reflective consciousness. For Merleau-Ponty, subject and object are *chiasmic* rather than atomic terms.[10] Although this revelation is positive for our present groundwork insofar as it foregrounds a richer range of meaningful phenomena than might otherwise show up under naturalism's objectivistic spotlight, we must nevertheless emphasize that *our own* ontological and epistemological investigations and contentions cannot be wholly extracted from the Gestalt structures that we, situated human theorists, inhabit. While, as already noted, this admission needn't entirely foreclose the more-than-human world from our investigations, it does mean that—apparently unlike the mainstream natural scientist—the ecophenomenologist cannot unproblematically take the meanings revealed in *their* phenomenology to be representative of the rest of the world. Merleau-Ponty illustrates this point with the secret "sexual pantomime" (*PP*, 184) of the dog or praying mantis, which a veterinarian, for example, might understand somewhat, but which doesn't show up for them in the *same* meaningful ways it does for those creatures. Each differently situated party is here invested in somewhat divergent Gestalts and so cannot address identical intentional objects at any level of description.

Thus, it seems that, if we're to be critically self-reflexive about our conceptual colonialism, then any ecophenomenological groundwork must concern itself with questioning the commitments and assumptions which generate the meanings we commonly perceive. Furthermore, given their irreducibility, our metatheoretical task must be conducted from within the confines of the sorts of Gestalt structures available to human body-subjects. Nonetheless, I hope to show over the remainder of the chapter that this insoluble tension between the body-subject's immanent participation within a Gestalt structure and her attempt to identify or narrate meanings which resonate beyond remains fruitful insofar as it forces a *radical* critical self-reflexivity about our imposition of conceptual violence. It is the sort of critical self-reflexivity that the aforementioned tension necessitates that may serve to disrupt the colonial mindset characteristic of our submerged mass. This process can only begin, however, with a phenomenological exploration of how meanings emerge within human perception. It is to this subject that we shall now turn.

1.2.2 Human Perception

We've seen Merleau-Ponty argue that perception is fundamentally intentional. He follows Husserl in distinguishing two types of intentionality in human perception: "intentionality of act, which is that of our judgements and of those occasions where we voluntarily take up a position"; and "operative intentionality . . . which produces the natural and antepredicative unity of the world and our life, being apparent in our desires, our evaluations and in the landscape we see" (*PP*, xviii).[11] The latter is the "condition of the former's possibility" (*PP*, 429). Act-intentional judgments instantiate the world's "objective" shape, but these reflective judgments aren't "free" in the Cartesian sense; they are normatively constrained by meanings generated through one's operative involvement within the world. Carl B. Sachs notes that, for the human as much as for the dung beetle, then, operative intentionality lacks "aboutness: there is no distinct intentional *object* even notionally separable from the intentional *act* directed towards that object" (2014, 105). While a body-subject's functional or teleological orientation cannot be overlooked without distorting the phenomena available for expression, a body-subject needn't aim at something *over and above* the way it reconfigures its leg movements, for instance. Even in the beetle's case, however, operative intentionality maintains "directedness; it is the intentionality of purposive behaviour" (Sachs 2014, 105). Since Merleau-Ponty refuses to bifurcate intentionality and consciousness, operative intentionality *just is* prereflective consciousness.

If operative intentionality is generative of our worlds and our subsequent reflective accounts, one might wonder how the meanings that delineate them are operatively garnered. Merleau-Ponty argues that bodily "motility" is "basic intentionality" in the sense that "consciousness is in the first place not a matter of 'I think that' but of 'I can'" (*PP*, 137). Like beetles or ticks, our perceptions of the world are intimately related to our bodily structures and the ways we typically comport them into it. Our bodies provide the normative background for any ontological and epistemological claims because entities gain their character from the intentional meanings they reveal within a phenomenal field. As Gestalt experiments show, things are differentially colored, deep or shallow, ordered or disordered, or inviting or repulsive in relation to our "our ways of inhabiting the world, and such inhabitation is always bodily in nature" (Cerbone 2008, 128–29). Since our bodily norms and idiosyncrasies are structurally essential to our "being-in-the-world," for any Gestalt, "one's own body is always the third term, always tacitly understood, in the figure-background structure" (*PP*, 78, 101).

Merleau-Ponty emphasizes that the most important role is here played by what he calls the "*body schema*" (*PP*, 98, translation altered).[12] Taylor Carman describes the body schema as "the bodily skills and capacities that shape our awareness of objects"; it is through the body schema that "intentional content" is "sketched out in advance by the dispositions that allow things to appear to us as they do" (2005, 68–70). Crucially, as David Cerbone explains, through the body schema, "the integrity of perception is informed by, and founded on, the integrity of bodily self-experience" (2008, 128). Take Merleau-Ponty's example of a marble rolled between thumb and forefinger. The individual digits connect with the object at points "that in purely sensory terms have nothing to do with one another" (Morris 2008, 116–17). Merleau-Ponty argues, however, that the intentional relationship of the movement of both digits to the marble's *anticipated* rolling affects the marble's synthesis as a unified object. The body schema thereby simultaneously reveals "the synthesis of one's own body, it is the reply correlative to [the synthesis of the object], and it is literally the same thing to perceive one single marble, and to use two fingers as one single organ" (*PP*, 205). Since the body more or less "presents itself as an expressive unity which we can learn to know only by taking it up, this structure will be passed on to the sensible world" (*PP*, 205). The body schema is, therefore, pivotal in dictating the diacritical relationships which establish the meaningful Gestalt-objects in which perception terminates. "The theory of the body schema is," therefore, "implicitly, a theory of perception" (*PP*, 206) since it discloses the primary mode of intentionality through which world and self are *expressively* revealed.

Nonetheless, the foregoing says nothing substantive about what it means for something to be "true" in an operative context. Merleau-Ponty seeks to explain this notion with reference to how, in perceptual acts, we shift our bodies with the operative goal of gaining maximal grasp upon the world: "My body is geared onto the world when my perception presents me with a spectacle as varied and as clearly articulated as possible, and when my motor intentions, as they unfold, receive the responses they expect from the world. This maximum distinctness in perception and action points clearly to a perceptual *ground,* a basis of my life, a general setting in which my body can co-exist with the world" (*PP*, 250, 267).

In Merleau-Ponty's account, maximal grasp is established through an embodied relation with worldly meanings that require minimal strain or instability on the part of the body-subject. Our embodied grasp of phenomena thus provides unavoidable normative impetus for the reflective distinctions between "truth" and "falsehood" that we subsequently draw. As Merleau-Ponty's shipwreck example shows, however, perception doesn't thereby become pejoratively "subjective," because there remain "success conditions; one can always misperceive by failing to adhere to the relevant perceptual norm" (Sachs 2014, 109). For Merleau-Ponty, unlike his naturalistic rivals, however, "success" cannot refer to whether one's "subjective" representations correspond to the "objects-in-themselves." Merleau-Ponty's success conditions are *immanent* to the Gestalt in which those phenomena are expressed. In the shipwreck example, for instance, the phenomena were truly determined only once the whole ship-object was instituted and each of its meaningful elements could more or less wholly satisfy the body-subject's operative expectations.

1.2.3 Sedimentation and Perception

If this were the whole story, however, we'd have difficulty accounting for the extent of the objectivistic prejudice manifested in our submerged mass. We've seen Merleau-Ponty credit the experience error with perpetuating objective thought. Nevertheless, given what he says about our irreducibly embodied grasp of the world's meanings, it seems unsatisfactory to attribute *the* source of our more markedly hierarchical or scientistic prejudices to a basic psychological error. While Merleau-Ponty provides compelling reason to link the experience error to our problematic preoccupation with the modality of the in-itself, it alone cannot explain why we take nonhuman things to be ontologically *reducible* to the terms of fundamental physics, for example. If, as embodied, one is as intimately bound up with the world's meaningful fabric as Merleau-Ponty contends, then the experience error provides an unsatisfactory explanation of the dominance of a scientistic purview which more or less denies the reality of those meanings outright.

To make sense of these prejudices we need to appreciate that, for Merleau-Ponty, meaning generated through the body schema is influenced by the intentional *sediment* of one's situation. Through the body schema, Merleau-Ponty argues, consciousness is "subtended by an 'intentional arc' which projects round us our past, our future, our human setting, our physical, ideological and moral situation, or rather which results in our being situated in all these respects. It is this intentional arc which brings about the unity of the senses, of intelligence, of sensibility and motility. And it is this which 'goes limp' in illness" (*PP*, 136). Through his appeal to the intentional arc, Merleau-Ponty asks us to reconfigure how we—under naturalism's influence—ordinarily think about the bodily abilities that shape the intentional objects we grasp. Although the body schema affords perceptual syntheses of body and world, it isn't a closed economy (as proponents of the reflex arc contend). For Merleau-Ponty, "habit expresses our power of dilating our being-in-the-world, or changing our existence by appropriating fresh instruments" (*PP*, 143). The intentional arc develops as experiential "knowledge" from one's multifaceted situation sediments into unreflective bodily habits. These habits then "enrich and [normatively] recast the body schema" (*PP*, 153). The line between operative and act intentionality becomes somewhat blurred here, because there is bidirectional transfer of influence between the two. In this sense, "every habit is both motor and perceptual, because it lies . . . between explicit perception and actual movement, in the basic function which sets boundaries to our field of vision and our field of action" (*PP*, 152). One's embodied *expectations* change in relation to one's goals, cultural history, social situation, and intersubjective experiences, and the habits in which these expectations are crystallized shape the meaningful physiognomy of entities encountered. As Merleau-Ponty puts it: "The gaze gets more or less from things according to the way in which it questions them, ranges over or dwells on them" (*PP*, 153), and this will depend partly on which instruments it operatively acquires.

This suggestion isn't as strange as it may seem. Stefano Micali (2013, 205–7), for example, notes that, despite possessing well-functioning physiological capacities, patients suffering from severe depression frequently report experiencing worlds which lack ostensive geometric depth, and which possess curious olfactory qualities. They usually report either that they experience a reduced capacity to affectively "feel" or distinguish smells and tastes or that they experience pervasive (usually repulsive) odors emanating from their bodies or surrounding world. What is particularly interesting about these cases is that what alters the meanings habitually located within the world is not most fundamentally a change in one's basic physiological structures, but a shift in how one relates to one's body and comports oneself into the world. For these patients, the aforementioned meanings appear to be intimately related to a phenomenal distance they come to experience from their own bodies—perceived here more significantly as an "object" or "burden"—which disconnects them from affective resonances previously afforded by the world and Others therein.

Sediment, however, is more all-encompassing than Micali's examples imply. We've already explored some banal cases where one's thematic experience shapes the meanings available at an operative level. Our veterinarian may identify a "canine sexual pantomime" only because the sediment of his training attunes him to the intentional salience of certain fine-grained movements, for instance. More importantly, as Sara Heinämaa emphasizes, for Merleau-Ponty, meaning is also the product of the sediment of our "intentional ancestors" in the sense that—insofar as one learns to perceive at all—one *always* takes up "an entire *tradition* of sensing and perceiving" (2009, 282). "My act of perception," Merleau-Ponty writes, "takes advantage of work already done, of a general synthesis constituted once and for all, and this is what I mean when I say that I perceive with my body or my senses, since my body and my senses are precisely that familiarity with the world born of habit, that implicit or sedimentary body of knowledge. . . . The person who perceives is not spread out before himself as a consciousness must be; he has a historical density, he takes up a perceptual tradition and is faced with a present" (*PP*, 238). These shared traditions, which we take up in their historicity, are essential to the intersubjective assent relations which help shape one's grasp of the "real." Why? Because one cannot take up reality on one's own; it is in the *intersubjective* world that "perspectives blend, perceptions confirm each other, a meaning emerges" (*PP*, xix).

Another illustration will prove helpful here. Linda Martín Alcoff (2006, 202–3) draws upon the psychologist Lawrence Hirschfeld's research, suggesting that, when learning to perceive race, children are directed to emphasize certain types of visible features over others. Antecedent ontological commitments about "natural kinds," therefore, largely determine which perceptual phenomena are ontologically realized as racial kinds. Children literally (though tacitly) *learn to see* certain meanings—here: contested racial ones—as historically situated, intersubjective norms dictate. While the children Hirschfeld studies *really do* identify some ontologically real phenomena, their meanings are disambiguated largely by the sedimented social and political structures to which they are subjected. Moreover, since, for Hirschfeld's subjects, an oppressive racial politics *precedes* their capacity to grasp the "objective" phenomena in question, this problematic sort of sediment isn't wholly separable from the child's grasp of the world at all. More worryingly, the politically charged meanings made intersubjectively available also inform or dictate "appropriate modes of conduct" (2006, 186) with members of the various naturalized racial kinds in ways that we, as irreducibly situated theorists ourselves, aren't always well attuned to identify or problematize.

It isn't hard to see how related sediment in the intentional arc might bear upon our objectifying habits and the violent behaviors they may license. The terms under which natural science subsumes the more-than-human world might be thought to be "naïve" and "dishonest" like the natural kinds mentioned above insofar as they neglect or dismiss the practical and intersubjective entanglements which influence scientific "objective" phenomena (*PP,* ix). We explored some contentions to this extent with Merleau-Ponty and Heidegger. In both cases, this naivety or dishonesty is further problematic because the *metatheoretical* reification or naturalization of certain existential meanings (at the expense of others) isn't obviously up for discussion in any revision of the theoretical boundaries which already rest, to some degree, upon those antecedent ontological commitments.

While an appeal to sediment might contribute to an explanation of our more puzzlingly reductive or scientistic objectivistic predilections, the question is: What kind of sediment could serve to inculcate them? Husserl (1970) rightly suggests that natural science's track record of predictive successes is an important reason why the scientific method has become equated with rationality in a contemporary context. Brown (2003, 7) also rightly claims that the reason that we frequently equate the extensional realist apparatuses of contemporary science with the a priori demands of ontology have something to do with the categories Descartes inaugurated. Evernden (1993) and Merleau-Ponty (*PP*, 72f) make related (and credible) points about our economic histories. Nonetheless, to my mind, the explanations which predominate the (eco)phenomenological literature lack some explanatory power regarding what exactly has continued to immunize a particular set of objectivistic metatheoretical assumptions against any significant revision over the past four hundred years. As I noted in the introduction, however, ecofeminists like Plumwood may fare better on this score. Since this is the main topic of the next chapter, however, I shall defer further discussion of it until then.

1.3 RADICAL REFLECTION

This concession notwithstanding, by attempting to identify and problematize objectivistic sediment, ecophenomenologists expose a further issue: if reflection is inexorably tied to a historically situated standpoint then, in one sense, the question posed above becomes unanswerable without recourse to the sort of transcendence rendered implausible by acknowledgment of one's embodiment. Otherwise put: if our default naturalism is symptomatic (and not merely the cause) of the sedimented ways that we objectify the world at the level of embodied habit, then objective thought isn't a straightforward cognitive error; it infuses the ways we're able to relate to the more-than-human world at an operative level. If, however, objective thought is more fundamentally an error of *praxis*, then it—and, by extension, the submerged mass of our crisis situation—becomes significantly more difficult to address.

If this is the context in which we find ourselves, then our ecophenomenological exploration of our submerged mass cannot be a wholly transcendental one either. If objectivistic sediment permeates our operative bond with the world, it is no longer unproblematic to simply usurp our default naturalism and ascribe to phenomena the meanings given in *our* phenomenological experiences instead. A naïve phenomenology of this ilk would fail to fully address the role of sediment in establishing one's grasp of the world and any meanings therein. Thus, if the violence characteristic of our submerged mass is largely precipitated, as I've suggested, by a failure of critical self-reflexivity, then even a transcendental ecophenomenology is somewhat problematically implicated in it. Our task, then, cannot be to bracket all worldly engagements, but to attempt to slacken "the intentional threads which attach us to the world and [bring] them to our notice" (*PP*, xiii). It is ultimately in taking up this somewhat paradoxical task of *radical reflection* that Merleau-Pontian ecophenomenology will prove particularly useful in disrupting the violence of the environmental crisis.

1.3.1 The World in Reflection

The issue of sediment problematizes transcendental phenomenology. But all is not necessarily lost. For one thing, Merleau-Ponty insists that, as embodied, one's grasp of the world cannot be reduced to a (subjective) representation *of* it. One's investigation of the meanings revealed in experience needn't also, therefore, be *wholly* constrained by sediment in the hopelessly constructivist manner that Cronon might insist on.

To explain, it's important to emphasize that, as Merleau-Ponty's marble illustrates, the world's impetus is essential to dialogically express perceptual phenomena and establish our bodies as unities in their own right. The marble's physiognomy wasn't *wholly* constituted by the body-subject, because the meanings available for expression were both licensed and limited by the world itself. To think otherwise would be to deny the world's own normativity, thus repeating the intellectualist's failure to provide a plausible account of the association of objects. For Merleau-Ponty, "the world is not what I think but what I live through. I am open to the world, I have no doubt that

I am in communication with it, but I do not possess it; it is inexhaustible" (*PP*, xvi–xvii). The grounds for communication become possible because, as body-subjects who are simultaneously both subject (*Leib*) *and* object (*Körper*), we're ontologically continuous with the rest of the world, which is "the setting of our own life, or our interlocutor in a sort of dialogue" (*PP*, 283, 320).

None of which is to say that one's grasp of marbles, for instance, cannot shift with the sediment of one's situation. One might become a marble connoisseur and identify previously undetectable imperfections within them, for instance. But this isn't Merleau-Ponty's point. Our ability to grasp objects *at all* depends on the basic intentional structures we *co*construct with the world. In this sense, Merleau-Ponty thinks, "radical reflection amounts to a consciousness of its own dependence on an unreflective life which is its initial situation, unchanging, given once and for all" (*PP*, xiv). Since, Toadvine argues, the "body's powers of expression are derivative from those of nature," this "gives a new twist to the sense in which the body 'sings the world'" (2009, 60; *PP*, 187). The first strand of radical reflection, therefore, emphasizes that one's reflective dependency upon prereflective experience must be, simultaneously, a dependency upon the world with which one is "connatural" (*PP*, 217).[13]

This is important. Although reflective consciousness reflects the sediment of one's lifeworld and objectifies in a manner not straightforwardly given at the operative level, it can never entirely break from the world's own meanings as constructivists sometimes insist. To reiterate: this is because the operative intentional bond upon which consciousness relies is, at base, conditioned *and* sustained by the world. There is, therefore, no reason to think that the meanings which show up for us phenomenologically wholly misrepresent those licensed by the more-than-human world itself, nor that the success conditions immanent to the Gestalt can be of no assistance whatsoever in uncovering unwarranted or misleading sediment. There are, after all, qualitatively important distinctions available *within* perception that justify the primacy of some meanings over others. Merleau-Ponty notes that "I cannot say that *I* see the blue of the sky in the same sense in which I say that I understand a book or again in which I decide to devote my life to mathematics. . . . I can see blue because I am sensitive to colours, whereas . . . I am a mathematician because I have decided to be one" (*PP*, 215). The sky's *anonymous* phenomenal blueness is something over which (unlike my reduction of a rainforest to lumber or energy flows) *I* have no real control—not because this perception isn't *mine*, but because the sky's blueness appears *irresistibly* licensed by the world itself. Its phenomenological primacy is such that it doesn't feel open to revision, no matter how intensively I interrogate it.

But this phenomenological primacy needn't be limited to such simple meanings. Since, as Brown argues, our everyday experience involves "moral sentiments that appear from a phenomenological perspective as instances of a prereflective axiological consciousness" (2003, 10), independent of prior objectivistic commitments, there is no obvious reason that *these* meanings must be less real than the sky's blueness. Indeed, insofar as the demand for radical reflection in our present context is explicitly motivated by "the sheer and simple fact of being struck by something wrong happening in the environment," one may *already* be sympathetic to Brown's contention. For him, "value experiences may be analyzed as a form of intentional consciousness in which the phenomenon of valuing and something valued are given together" (2003, 11). Just as anger may be noninferentially present in an expression, and love may be noninferentially present in a gift, where we suspend our sedimented intentional commitments sufficiently to be open to it, we might perceive the callousness, greed, or wrongness noninferentially manifested *within* concrete instances of deforestation or oceanic plastic pollution (*PP*, 186, 321).

It would be a mistake, however, to invoke a straightforward *reversibility* between those meanings latent within the more-than-human world and those expressed by human body-subjects. Although prereflective dialogue needn't navigate the distorting membrane of reflective consciousness, we've seen reason to believe that it is *always* colored by habitual sediment, even at the operative level. Here is our dilemma: On the one hand, reflection manipulates the world's meanings "because the representations of such thinking inevitably change our basic experience by introducing categories and conceptual distinctions that were not originally given there" (Shusterman 2005, 167). It was the manifestation of this problem within objective thought which motivated a shift toward phenomenological methodology. On the other hand, however, the operative dialectic essential to express the world's meanings is so vulnerable to misleading sediment that to fail to reflect would be deeply problematic. If we're to attempt to return to the things themselves in earnest, then we need recourse to a second strand of radical reflection.

1.3.2 Reflection in the World

The first strand highlights the impossibility of establishing reflective conclusions that entirely transcend the intentional bond upon which expression relies. Our second strand is also concerned with the situation taken up in reflection, but with a reverse focus on the role of sedimentation and reflective distortion in establishing prereflective meanings. This strand is principally defined by Merleau-Ponty's contention that reflection "is truly reflection only if it is not carried outside itself, only if it knows itself as reflection-on-an-unreflective-experience, and consequently as a change in structure of our existence" (*PP*, 62). I take this to mean two things: firstly, that our operative grasp—to which reflection is tied—is necessarily manipulated by that thematic reflection; secondly, that in being carried out within itself, reflection will be shot through with the sediment of one's multifaceted situation. Reflection may become *truly* radical only when, as it reflects on its underlying phenomena, it addresses both concerns in its second-order reflection upon itself.

How is this possible, given Merleau-Ponty's warnings that "none of [one's] thoughts will be able to be quite detached from the historical context in which [one] lives," and that "to be situated within a certain point of view necessarily involves not seeing that point of view" (*PP*, 72f; *SB*, 217)? We can elicit from Merleau-Ponty's work a sense in which the real *radicalness* comes through embracing the somewhat paradoxical task taken up in an engaged and perpetual attempt to return to the more-than-human things themselves, without hope of success. Our second strand of radical reflection takes entirely seriously that, especially given our inescapably intersubjective means of grasping the world, one cannot hope to reflectively return to, or coincide with, any absolutely presedimented, prereflective worldly bond. *That* is "a past which has never been a present" (*PP*, 242). Instead, Merleau-Ponty argues, "the most important lesson which the reduction teaches us is the impossibility of a complete reduction" (*PP*, xiv). A transcendental ecophenomenology of the early Husserlian ilk *is* impossible. But through our situated attempts to perpetually investigate the objectivistic and utilitarian residue of our situation, we may inculcate a more positive *existential* phenomenology. In this existential ecophenomenological context, meanings—including any axiological ones—are uncovered only hyperdialectically; they are "subject to continual reassessment in light of our subsequent experience, just as we continually reassess our previous understandings of the Real or the True" (Brown 2003, 11) within a critical social praxis.[14]

We should be reminded that genuine *progress* becomes possible in radical reflection because of the intentional bond described above. Nonetheless, in this existential ecophenomenological context, "perception is not presumed true, but defined as access to the truth," and any provisional claims about the "real" are beholden to one's "perception of the world upon which our idea of truth is forever based" (*PP*, xvi). While we should not and cannot abandon our tacit affirmation of the world, we must retain a perpetual and radical sort of critical self-reflexivity about even the most basic meanings we take the world to license. In the present environmental context, then, I take real progress to arise, not solely from the metatheoretical revision of "subject" and "object" terms which Merleau-Pontian ecophenomenology forces, but also from the more fundamental change in *how we relate to the world* which an appreciation of how those terms arise in our phenomenologies demands. I'm suggesting, in short, that a concerted attempt to address our submerged mass must come through a shift from objectification to radical reflection, *as a matter of praxis.*

So, what does this mean for the environmental crisis specifically? No matter how skilled an ecophenomenologist, one cannot straightforwardly or unproblematically uncover *the* fundamental or univocal presedimentary meanings by which environmental "issues" (e.g., depleting rainforests, polluted oceans, or collapsing ecosystems) should be understood. Nor will radical reflection itself straightforwardly reveal *the* (economic, technoscientific, or political) policies or maxims with which to "solve" them. But, understood in an existential ecophenomenological context, this just *isn't* the primary role of the environmental philosopher. In fact, the demand that environmental philosophy *should* itself aim to achieve one or more of these goals is symptomatic of the totalizing philosophical tide against which, insofar as they are radically reflective, ecophenomenologists must swim.

What ecophenomenological philosophy can do, however, is to help inculcate a certain kind of *humility* by disrupting the totalizing, utilitarian, and objectivistic thinking which appears to precipitate those "issues" in the first place. By taking up something akin to what Husserl called a "*habitus* of critique" (Husserl, quoted in Miettinen 2013, 341), we might begin to disrupt our colonial mindset and, thus, make progress in addressing the underlying violence of our crisis situation. Taken in this sense, environmental "philosophy does not consist of informing individuals on 'what to do' or 'how to live'—rather . . . its task is that of motivation, the constant calling forth of critical self-inspection" (Miettinen 2013, 341). And that is why the critical environmental philosopher must be a "a place of unrest." Unlike in the Hegelian usage, however, there can be no hope of absolute synthesis without undermining the productive tension inherent to the paradox. "Hegel," Merleau-Ponty tells us, "is the museum" (*ILVS*, 82).

1.4 CONCLUSIONS

This chapter has begun the groundwork for those to come. I began by challenging natural science's onto-epistemological monopoly on the terms under which the more-than-human world is characterized. I suggested that naturalism's underpinning objectivistic assumptions and commitments distort the phenomenological data on which natural science is based, and that these assumptions conceal problematically utilitarian biases which license colonial ways of relating to the more-than-human world. Since these objectivistic purviews are directly implicated in the violence characteristic of our environmental crisis, I argued, natural science should play a more limited role in determining the terms under which we understand and attempt to respond to it.

Taking the phenomenological underpinnings of our ontological and epistemological contentions more seriously, I argued, requires an attempt to return to the perceptual phenomena themselves. Over the remainder of the chapter, we explored the promise and limitations of Merleau-Pontian ecophenomenology's ability to do so. We saw that natural science's conceptual violence derives, in part, from its tendency to mistake the Gestalt-objects in which perception concludes for the basic constituents of ontological reality. Through closer attention to our multifaceted Gestalt-situatedness, that Merleau-Pontian ecophenomenology might address this violence by both licensing a richer range of ontologically real phenomena than can be registered naturalistically, and by encouraging habitual critique of the ontological and epistemological terms we extend to more-than-human entities. Although, given our intractable immanence within the world investigated, radical reflection is somewhat paradoxical, I've suggested that this paradox may prove fruitful in the present *eco*phenomenological context insofar as it licenses a certain humility via a shift in habitual praxis away from uncritical objectification. Over the following chapters, I will explore some more concrete suggestions about how such an ecophenomenological praxis *might* look.

CHAPTER TWO

ECOFEMINISM AND ECOPHENOMENOLOGY

In the previous chapter, I argued that ecophenomenology has some advantages over our default scientific naturalism because natural science's underpinning metaphysical terms are reflective distortions of those given in our grounding phenomenology. By rejecting naturalistic ontological presuppositions, Merleau-Pontian ecophenomenology may acknowledge the epistemic salience of its own diacritical background and remain radically reflective about its background commitments as a matter of praxis. Thus, I aimed to show, Merleau-Pontian ecophenomenology is amenable to doing less conceptual violence to more-than-human entities than its naturalistic rivals. Since the behavioral violence which precipitates our crisis situation appears to be intimately related to the conceptual violence which a Merleau-Pontian approach may help alleviate, a Merleau-Pontian ecophenomenological praxis shows significant promise for our task.

In this chapter, I explore powerful intertwinings between ecophenomenology and ecofeminism which support the role I claim for ecophenomenology. After outlining Val Plumwood's ecofeminism, I first contend that ecofeminist insights can help ecophenomenologists flesh out some of the unwarranted sediment implicit in objectivistic conceptions of the more-than-human world. Ecofeminists, I argue, help us further appreciate the dangerous falsity of taking objectivistic subject/object divisions to be metaphysically primitive by showing that these divisions are *politically* motivated in a manner that is set *specifically* against nonexploitative engagement with nonhuman entities. Nonetheless, I also argue that ecofeminist insights benefit from being reconfigured ecophenomenologically to address the curious lack of attention to embodiment which, contrary to their official commitments, infuse the ecofeminist accounts I address. I finally argue that, by addressing dualistic assumptions as more-than-cognitive errors, a Merleau-Pontian ecophenomenological praxis has the potential to more effectively challenge the logic of domination that ecofeminists diagnose in objectivistic modes of revealing. Since ecofeminist insights may be successfully incorporated into an existential ecophenomenological framework, I argue, this further supports adopting an ecophenomenological praxis to attend to the "submerged mass" of our crisis situation.

2.1 ECOFEMINISM

Although the term "ecofeminism" was probably coined by Françoise d'Eaubonne in 1974, its roots go much deeper. Simone de Beauvoir, for example, claimed in 1949 that a common patriarchal logic sets women and nonhuman entities apart as pejoratively "other" (Glazebrook 2002, 12; Beauvoir 1952, 114).[1] Ecofeminism is also a broad church. As I use the term, "ecofeminism" designates theoretical approaches which diagnose *nonaccidental* connections between the domination or exploitation of women, and of nonhuman entities. The specific variant of ecofeminism I'm concerned with here contends that exploitative behavioral patterns are underpinned by a common (but contingent) conceptual *logic,* which is overlooked or neutralized in ordinary discourse. Ecofeminism should be of interest to environmental theorists because, if ecofeminist arguments are successful, key aspects of our environmental crisis ranging from anthropogenic climate change and overpopulation to deforestation and water depletion result partly from the normative logic in which our conceptions of the more-than-human world are steeped. Addressing dualistic logic wholesale will, therefore, help us address the crisis at root. Although forms of this narrower claim are common to most ecofeminist theories, I shall now outline Val Plumwood's contribution since she offers arguably the most compelling version of it.

2.1.1 Plumwood's Ecofeminism

Plumwood (1993; 2002) argues that dualistic logic is so dominant, at least in "Western" societies, that its ubiquity grants it a sort of invisibility as naturalized discourse.[2] By "dualism," Plumwood doesn't mean to refer to the simple distinctions between conceptual pairs which may be understood to some extent independently, and which we couldn't do without in making sense of environmental theory (e.g., self-other or public-private). Dualisms exist where one side of the pair affirms its superiority over, and—most importantly—denial of dependency upon, the other. Dualistic logic thereby entrenches a value-laden master/slave relationship which shapes *both* relata. Given that both "slave" and "master" identities emerge only in the context of a colonial relationship, dualistic logic is, Plumwood argues, inescapably tied to domination (1993, 47). Under dualism, naturalized domination relations therefore delineate our accounts of ontological reality, our identities, the axiologically justified order of things, and so on.

Furthermore, Plumwood (2002) argues, since "Western" culture is unequivocally driven by rationality, answers to pressing environmental "issues" (What should we do about anthropogenic climate change? Should we value biodiversity? How should we mitigate habitat depletion?) are couched in rationalistic terms. The selection of what is considered "rational," however, is underwritten by the same self-perpetuating, ideological schema. "Solutions" to environmental "issues," therefore, share the familiar problem of being couched in the same terms as those partly responsible for their arising in the first place. Genuinely liberating solutions can, Plumwood thinks, only arise by unseating the entire dualistic system and the domination relations essential to it. Rationality will look very different afterward. Attending to this task is Plumwood's contribution to addressing our "submerged mass."

2.1.2 Dualisms

According to Plumwood, a multitude of dualisms form a mutually reinforcing, "interrelated set" (1993, 42), which collectively denigrate their underside. However, I will focus on only four: "subject/object," "reason/nature," "culture/nature," and "male/female" (1993, 43). To briefly illustrate how these dualisms operate, consider the relatively uncontroversial observation that an enduring source of women's oppression derives from their identification with "mere" (i.e., somehow *non*human) nature. Such inherently pejorative identifications usually appeal to a common "animality" or reduction to immanence and, thus, allegedly limited intelligence or capacity for self-control. As Plumwood notes (1993, 19), Plato, Bacon, Descartes, Hegel, and Burke all make this link more or less explicitly. But there is something even more sinister than patriarchy afoot. By being *defined* antithetically to reason, culture, and other valuable aspects of humanity, "Nature" here becomes a *political* category.[3] Dissociation of the valuable master side from "mere nature" also then motivates the thought that the underside is something to be *overcome*.

Here are some examples to substantiate Plumwood's claim: Plato (in)famously thought the mutable, corporeal world useless in attaining knowledge of the ultimate reality. In the *Republic* (1987), the bestial body (nature) must be rationally dominated if one is to attain knowledge. In Plato's *Timaeus*, the "feminine" is explicitly identified with the appetitive: the "lowest" form of oneself that must be suppressed if one is to flourish (Plato 1965, 71; Plumwood 1993, 91–93). Francis Bacon—the founder of the modern scientific method—and his successors (notably Leibniz) notoriously took up the popular conception of nonhuman nature and women being equally reducible to their reproductive capacities to speak of the former as a separable feminine *object*. For Bacon, progress requires that the scientific inquisitor (a wholly detached "man of reason") *force* "nature" to *yield* to his demands to "deliver" her truths—truths which, despite Bacon's use of violent terms like "vex" more commonly associated with torture, are allegedly passively received by their "impartial" epistemological interrogator (Merchant 1980, 172–78). A similar legacy is taken up in Cartesian conceptions of the *res extensa* ("object") investigated by the *res cogitans* ("subject"). In each case, "man" (as shorthand for humanity) is defined in opposition to (nonhuman or nonrational) "nature," with which woman is pejoratively identified. We must explore dualism's dynamics, however, to see how dualism survives in subtler forms.

2.1.3 Dualistic Dynamics

Backgrounding. According to Plumwood, dualisms are established through five interrelated dynamics which magnify or manipulate differences to reinforce the hierarchism which precipitates dualism's colonial logic. "Backgrounding" entirely denies the master's dependency upon the slave.[4] This denial is, however, grossly misleading. Although master identities can be separately conceived, they are (as in Alexandre Kojève's [1980] Hegel) conceptually defined only relative to their domination of the slave. Moreover, "the master requires the other materially, in order to survive, for the relation of complementation has made the master dependent on the slave for fulfilment of his needs" (Plumwood 1993, 49). Even Bacon's "man of reason" is embodied. His rationality, therefore, presupposes his carnality and the material relationships with the more-than-human world upon which his "hyperbolised" (Plumwood 2002, 17) autonomy depends. The backgrounded dependencies Plumwood has in mind are primarily appetitive or respiratory (there must be sufficient food to eat and air to breathe), but we might also include the Baconian interrogator's subjection to the "physical laws," mentioned in the previous chapter, which delineate "his" discoveries.

Backgrounding also has a social dimension. We've already explored how intersubjective relationships might shape our ontological and epistemological contentions in surreptitious ways. Ecofeminists, however, focus on how dualism provides normative political impetus to underplay social dependencies in ways deeply problematic for dualism's underside. Ecofeminists frequently observe that identifying women with "mere nature" ensures that women's contributions to "rational" and "cultural" achievements are characterized as passive or inessential. Socially reified "women's labor" (e.g., raising children or growing subsistence crops, rather than contributing directly to the formal economy), for instance, tends to be subsumed into the "nature" category and these essential agentic contributions to cultural accomplishments are obscured. Dualism thus marks both women and nonhuman nature as the Lockean *terra nullius* on which valuable cultural capital is built (Plumwood 1993, 4).

Here is a salient example. Trish Glazebrook (2011) argues that backgrounding has hampered Ghana's recovery from flooding linked to anthropogenic climate change. Under these conditions, food security is undermined and, since around 70 percent of the world's farmers are women (the percentage in Ghana is higher and most are subsistence farmers), the resultant problems affect women disproportionately. Tellingly, however, (inter-)governmental aid has focused on funding development strategies that elevate gross domestic product (GDP). Doing so effectively means displacing or failing to compensate subsistence farmers who lack a *directly* measurable impact on GDP. Even setting aside the dubious economic merit of these strategies, they present an obvious further issue. Women here are particularly vulnerable to negative effects of both the "issue" (anthropogenic climate change) and what purports to be the "solution" (technoscientific developmentalism), because of their place in the dualistic schema. The large-scale, mechanized monocultures pursued by even "green" developmentalism do systemic violence to women in a similar fashion (Glazebrook 2002, 17). Although a related "invisibility" for subsistence farmers permeates ongoing work in climate change mitigation and adaptation, however, the 2010 "UN Framework Convention and Kyoto Protocol [still did] not identify women as a vulnerable group or as crucial actors in mitigation and adaptation efforts" (Glazebrook 2011, 774).

Radical Exclusion. Backgrounding establishes a normative hierarchy. Through "radical exclusion," the importance of differences between parties is amplified or mythologized to "hyperseparate" master from slave, such that *he* is absolved of significant continuity with *her*. Dualism achieves this largely via a "Differential Imperative," which seeks to establish identity as a matter of "kind not degree" (Rodman, cited in Plumwood 1993, 71, 50–51). Identities are radically divorced, often based on one (perhaps insignificant or distorted) property. Historically, radical exclusion has focused on the capacity for reason or "subjective" consciousness, which, for Descartes, becomes *man's* essence. Radical exclusion may, however, operate more subtly to cleave master/slave orders. Women's identities are still more frequently tied to their reproductive capacities than men's, for example. Likewise, Greta Gaard notes, "in terms of radical exclusion, queers find that the erotic (a particularly perverse erotic) is projected onto queer sexuality to such a degree that this quality is seen as the only salient feature of queer identities" (1997, 116).

Interestingly, given her contentions about the futility of developing environmental rationality from within dualism, Plumwood also identifies "neo-Cartesian" (2002, 147) tendencies among contemporary animal ethicists. These are evident through the exclusionary thresholds which Tom Regan (minimal subjectivity) and Peter Singer (sentience), for example, draw for inclusion into the moral sphere.[5] Singer and Regan apparently err by extending moral considerability only to nonhumans who demonstrate sufficient psychical continuity with human consciousnesses, rather than problematizing the master/slave dynamic which legitimates any such radically exclusionary mechanism. Thus, Plumwood argues, Singer's and Regan's "solutions" to "issues" of animal welfare are of limited value in addressing the broader mindset responsible for our domination of the more-than-human world. For Plumwood, however, Singer's and Regan's limitations don't straightforwardly derive from an unerring commitment to naturalism. They derive from adherence to the same hierarchical *logic* which is apt to set "agent" and "resource," "mind" and "nature," or "subject" and "object" apart in kind.

To elaborate, imagine that, under Singer's influence, we (somehow!) establish a paradigm of near-global veganism. Doing so would probably have a positive impact on animal welfare, soil erosion, and carbon emissions. It would, however, leave remarkably untouched the more fundamentally utilitarian assumptions which legitimate enframing and relating to more-than-human entities *as* resources. Most obviously, this is because Singer refuses to grant nonsentient things (oysters, perhaps, but also abstract or composite entities like "biodiversity," rainforests, or specific marine ecosystems) moral considerability in their own right. But the same sort of dynamic (via the related mechanisms of "instrumentalism" and "homogenization" explored below) requires Singer to conceive of even sentient entities as ideally calculable aggregates of utility which should be *substituted* (i.e., function as resources) for greater ones. For Singer, a legitimate (or morally praiseworthy) substitution may take place when heavy-use road infrastructure projects win out over the survival of rare great crested newt habitats (Alaimo 2017).

Radical exclusion's faults don't end there. Since radical exclusion denies the master's manifold dependencies, it also demands a foundational atomism within our conceptual schemes. Dualism thereby implicitly naturalizes metaphysical presuppositions that, experimental evidence suggests, are generally favored by men (broadly, the empowered group), and which perpetuate themselves by skewing future ontological cross-referencing *against* recognition of relationality (Hintikka and Hintikka 1983, 146). By *beginning* from these antirelational assumptions, dualism even skews which entities natural scientists can find in the world. This is because, under dualism, natural scientists (and, subsequently, naturalist philosophies) are beholden to the same basic concepts of essence, causality, and agency which structured the early modern worldview (Bannon 2014, 35–36). Moreover, this purview demands that "minds" be cleaved from "nature." Radical exclusion, therefore, demands much the same reflective purification of "object" from "subject" problematized in the previous chapter.

By understanding "unbridgeable" (Plumwood 1993, 51) qualities like the above to be definitive of the natural order, radical exclusion legitimates anthropogenic exploitation for many of the same reasons that Heidegger's analysis of technology foregrounds. Dualism's metatheoretical dynamics and structural assumptions ultimately *naturalize* colonial attitudes and relationships through the hierarchical and utilitarian terms they permit. Like Heidegger, Plumwood argues that our ontological and epistemological contentions are as infected by these prejudices as our behaviours. But Plumwood also gives reason to be cautious of uncritically using the term "anthropocentrism" to explain this problem, because that term may itself court a problematically exclusionary logic. Plumwood inadvertently thereby highlights some concerns we might have about contemporary Anthropocene rhetoric if we're to address the environmental crisis that this "epoch" allegedly foregrounds. I explore these in the next chapter.

Incorporation. "Incorporation" further denigrates the underside by defining it *only* relative to the master. Incorporation makes the slave's characteristics appear derivative, rather than acknowledging that master and slave share an "equally relational definition" (Plumwood 1993, 52). The master perspective thereby gains an air of anonymity or universality such that its contingency on illicit conceptual origins is lost. The slave, however, is marked as "other," usually *lacking* one or more of the master's positive qualities. Incorporation thus ensures that, under dualism, one simply isn't free to relate to more-than-human entities as equals: "The other is recognised only to the extent that it is assimilated to the self, or incorporated into the self and its systems of desires and needs: only as colonised by the self" (Plumwood 1993, 52). Insofar as they admit nonhuman entities to the moral sphere solely via extension of qualities of humanoid experience, incorporative tendencies are as evident in the allegedly liberatory philosophies of Singer and Regan as in Kant or Christine M. Korsgaard's more straightforwardly anthropocentric visions.[6] As we shall see in the next chapter, qualities which problematize their respective placement in an incorporative schema are ordinarily backgrounded.

Instrumentalism. The final two features are important corollaries to radical exclusion and incorporation. The first is "instrumentalism," under which "the lower side is objectified, without ends of its own which demand consideration on their own account. Its ends are defined in terms of the master's ends" (Plumwood 1993, 53). Instrumentalism affords the master perspective a certain privileged "neutrality" which allows it to be uniquely valuable in its own right. To reprise an earlier example, instrumentalism helps explain why subsistence farming or children's education is valued largely only to the extent that it contributes to the formal or rationalistic "knowledge economy." Instrumentalism may also play a key role in licensing the ratiogenic violence done by reducing more-than-human entities to exchange-value under capitalism or standing reserve under technoscience.[7] Plumwood (2002, 16) goes further than Heidegger on this issue, however, by arguing that technoscience's lack of critical self-reflexivity about its ontological and epistemological presuppositions is, ironically, *irrational* in a greater sense. This is because the unsustainable paradigm of consumption which technoscience perpetuates is arguably the largest determinant of behaviors which threaten the master's very survival (e.g., where anthropogenic desertification makes the Earth inhospitable to growing grain). Although instrumentalism ordinarily presents master/slave relationships as somehow mutually beneficial, this indicates one way that dualism may clearly be bad for *both* of its relata.

We should nonetheless emphasize that, regarding the "environmental crisis," things are much worse for some than for others. The effects of anthropogenic climate change, for example, are suffered more acutely by groups marked as "other." Plumwood gives the example that people who cannot afford to migrate or install air conditioning are at much greater risk of dying when temperatures rise (as in the 1995 Chicago heat wave) than their more affluent contemporaries (2002, 84). Nancy Tuana (2007, 207–8) notes, via reflections on the unequal death toll of Hurricane Katrina, that cocausal relationships between disability and poverty further inflate economic and material barriers to migration. Gaard (2015, 23) notes that rural desertification in developing countries also affects women more significantly than men because, given the instrumentalism of "women's labor," women in these communities are more likely to starve so that male family members may eat. Indeed, The International Union for Conservation of Nature estimates that "women and children are 14 times more likely to die in ecological disasters than men" (Gaard 2015, 23). The reality behind this statistic suggests an intimate link to dualistic logic. Women are often confined to households; they find it difficult to escape, given the burden of carrying children (which men do not typically share); they are reluctant to leave because of the (statistically well-justified) fear of being raped; and so on. Examples abound.

Homogenization. "Homogenization" further denigrates the underside by refusing to recognize salient differences within any group that "deviates" from the occupied center. The underside are particularly vulnerable to homogenization, Plumwood thinks, "because they lack the power to require recognition in their diversity" (2002, 103). Homogenization also licenses substitutability by denying the particularity of members of a "universal" class, thereby facilitating the master/other dichotomy essential to the other mechanisms (Plumwood 1993, 53–54). To illustrate, Plumwood (1993, 54) points to Marilyn Frye's contention that the "natural order" of dominion gains plausibility by reifying two sharply demarcated sexes with homogenous features shared by *all* members. Even setting aside that this picture fails to track people who identify as transsexual and so on, the biological essentialism upon which it relies is problematized by the available scientific data. A study by Brown University, for instance, was unable to estimate the frequency of intersex births because the result depended on what one took to be the biological marker(s) of human sex (Blackless et al. 2000). The study proposed seventeen potential categories, in which the estimated frequency ranged from one in 100 to one in 150,000 births. The nonhuman picture is at least as ambiguous. In solely genetic terms, some species of fungi have over 36,000 sexes (Arthur 1999).

Homogenization is also more subtly manifested in eco-holistic environmental philosophies like John Baird Callicott's. Guided by ecological science,

Callicott (1989, 109) understands ecosystemic entities to be fundamentally reducible to energy flows. For Callicott, "objects" (including individual animals) possess less ontological reality than their "anonymous" components, which, for Callicott, are exactly the sort of ideally calculable energy reserves already problematized. By establishing this ontological hierarchy, Karen J. Warren (2000, 155) argues, Callicott licenses the in-principle substitutability of *all* macrolevel entities (e.g., subsystem, wolf cub, badger cete) for the sake of the composite ecosystem they form. We should, however, emphasize that Warren's objection isn't *only* that Callicott's ontological and axiological contentions are restricted by the problematically reductive and utilitarian terms he inherits from ecological science. Nor (I suspect, wrongly) does Warren think Callicott's naturalism problematic in requiring him to address the modality of the in-itself. Her main objection is that the scientific determinations which Callicott inherits and their subsequent axiological extensions are determined not, most fundamentally, by the empirical evidence, but by the colonial *political* dynamics of a dualistic conceptual scheme. In fact, I take *this* claim to exemplify the overall objection that ecofeminists pose to the ways we presently conceptualize environmental "issues" and "solutions": dualism provides the "common sense" basis which underpins *any* (scientific, political, or philosophical) understanding of the way things are and should be, but only in the Gramscian sense, where "common sense" means "sedimented hegemony" (Gramsci 1971).

The question of how to respond to dualism, however, is tricky. Historically, feminists have taken two equally unsatisfactory approaches. The first is to argue that the (unjustified) cleavage between men and women doesn't parallel the (justified) cleavage between, say, human culture and "mere nature." This approach was taken by liberal feminists like Mary Wollstonecraft (1995) and, arguably, Beauvoir (1952). The other approach is to embrace an essential differentiation which, some such theorists hold, better positions women to address environmental issues than their male counterparts. This approach was taken by early ecofeminists like Ariel K. Salleh (1984). One central problem with both approaches is that, by enacting relatively uncritical reversals of the dualistic schema, they risk colluding with the colonial logic which establishes master/slave identities in the first place.[8] By leaving dualistic logic significantly intact, uncritical reversal strategies remain hostile to nonexploitative relationships with the more-than-human world. Plumwood argues that addressing our submerged mass, therefore, requires wholescale disruption of the distortion of difference that demands an attitude of domination from the outset. In chapter 3, we shall explore how an ecophenomenology informed by ecofeminist insights might begin this task. But first, we should tie up some loose ends concerning the appropriate way to understand the relationship between these theoretical perspectives. We shall do so by first exploring some fruitful intertwinings between them.

2.2 INTERTWININGS

2.2.1 Objectivistic Sediment

The previous chapter focused on how our default scientific naturalism is problematic because it is inconsistent with its underlying phenomenological data and its justification rests upon objectivistic ontological assumptions that Merleau-Ponty tells compellingly against. Ecofeminism's first important contribution to our groundwork project is to allow us to better understand the sediment which precipitates these persistent objectivistic commitments.

I noted in the introduction that the tendency to wholly bifurcate subject and object can be traced back at least as far as modernity. However (unlike Bordo [1987]), Plumwood resists the temptation to make Descartes or his contemporaries the *inaugurators* of dualism.[9] One reason for doing so is that, by appreciating its logical and political precedent, we're better able to understand why the moderns subsumed the more-than-human world under the reductive terms they did. If the logic of dualism is inherently one of domination, then their mind/nature hyperseparation becomes more understandable. Dualism is, after all, facilitated by a polarizing dynamic set against appreciating one's physicality or material relationality from the outset. It is further plausible that Descartes and his successors' ability to "clearly and distinctly" imagine the substantial separation of mind (qua subject) and body (qua object) derives largely from the dualistic metatheoretical assumptions they inherit.[10]

Ecofeminists also thereby help explain why our objectivistic stipulations survive the formal demise of precritical metaphysics. I noted in the previous chapter that, while the experience error might motivate the naturalist's presumption of a world furnished with determinate objects or properties-in-themselves, it wouldn't straightforwardly legitimate the more reductive or utilitarian terms under which we take contemporary physics, for instance, to exhaust ontological (or at least empirical) reality. Plumwood (1993, 120), however, cashes out some of the salient sediment involved. She argues that the recurrent tendency to address the mind/body problem in abstraction from its sociohistorical context ensures that important metatheoretical assumptions are glossed over in the theoretical revisions they license. By neglecting the colonial logic which helps establish the metaphysical plausibility of Cartesian or Baconian "objects," subsequent revisions are, therefore, bound to inherit *political* assumptions that are dualistic like their forebears'. Where Plumwood's contention differs substantively from Brown's is in her suggestion that our ongoing failure to address the Cartesian precedent is partly motivated by a political undercurrent which surreptitiously informs what subsequent revisions *should* show.

Plumwood also thus helps clarify why the prevalent challenge to substance dualism in the philosophy of mind, for instance, comes not from quarters that challenge the essential properties by which sides are characterized, nor those that problematize hyperseparational logic, but through variants of physicalism.[11] Since physicalism more or less affirms one side of the subject/object dualism, however, Cartesian essentialism persists through extensional realist metaphysics.[12] This has the ostensibly positive result that natural science and philosophy may coincide in a manner which contradicts Stephen Hawking's barb that "philosophy is dead. Philosophy has not kept up with modern developments in science, particularly physics. Scientists have become the bearers of the torch of discovery in our quest for knowledge" (Hawking and Mlodinow 2010, 13). The problem with physicalism's "*truncated reversal*" (Plumwood 1993, 121) of the Cartesian schema, however, is that physicalism reestablishes a familiarly impoverished concept of the "physical," which prohibits nonexploitative means of relating to the more-than-human world through its utilitarian and instrumentalist terms.

Ecofeminism has further merit in explaining the political motivations for continuing to understand the metaphysics of scientific investigation as we do. I've argued that even Kuhnian natural science adheres to a somewhat Baconian paradigm under which a distinct epistemic subject investigates an ideally purifiable object. If my previous arguments hold, the scientist must fundamentally take himself to be (as Bacon claims) epistemically *passive* because even intellectualists require the receipt of objective stimuli to pass reflective judgment upon the world-*in-itself* (in at least an empirically real sense). However, the violent metaphors Bacon employs to describe scientific methodology reveal more than he intends here. This is because, like Merleau-Ponty and Heidegger, Plumwood provides reason to believe that a metaphysically purifiable order of "objects" can be maintained only by imposing restrictive and exclusionary experimental apparatuses and conditions upon the world. Under the logic of dualism, however, retaining Baconian subject/object relations makes perfect sense, since, in the name of domination—here taking the form of *totalizing* knowledge—the aforementioned logical dynamics legitimate the atomic independence of parties and, therefore, license the hyperseparation essential to the paradigm. If central ecofeminist arguments hold, however, the parameters of this paradigm reflect the interests or distorted (e.g., hyperbolically autonomous) identity of the master. Naturalism is thus led to court (as Merleau-Ponty also claims) a disembodied, but not disinterested, logocentrism.

Ultimately, for Plumwood, the subject/object dualism which effects the requisite metaphysical distance to license transcendent or transcendental accounts of the "objective world" amounts to an epistemic analogue of the person/property dualism (2002, 41). Positions which employ (truncated reversals of) it are, she claims, therefore predisposed to commodify the more-than-human world. This is because objectivistic schemas aren't *just* metaphysically problematic (as Merleau-Ponty shows); they are also politically problematic because, under dualism, "objects" are also simultaneously underside. This plays out within our dominant forms of naturalism in terms of a "subject" who may *possess* the "objective world" and what are, purportedly, "its" truths, due to the limited terms under which it is subsumed. If the "passification of the objectified is," as Plumwood (2002, 46) writes, "a prelude to their instrumentalisation," then attending to the violence of our submerged mass appears to require remedying our default naturalism and its underpinning dualistic sediment.

These conclusions might appear too quick. After all, as Karen Barad (2007, 125–26) notes, the representationalist separation of "objects" from mediated perception—and, thus, the reduced plausibility of their epistemological *possession*—is widespread among both realists and social constructivists in contemporary science and science studies debates. However, as Barad also notes, the only real ground for debate between scientific realism and constructivism is whether one's "subjective" representations may correspond with a distinct world-in-itself or are culturally determined. Both still hyperseparate human "culture" ("subjective" representation) from nonhuman "nature" (represented "object") in a manner that Merleau-Ponty and Plumwood think is a problematic reflective distortion. Both also remain dualistic in relying upon the same old *hierarchical* subject/object divisions under which the former is agentic and active and the latter inert and plastic. Moreover, given their common metaphysical presuppositions, realism and constructivism both ultimately slip back into the colonial mindset which underwrites the Baconian account: either the world-in-itself *can be exhausted* by the "appropriately attentive" observer, or else (as appears to be the case for William Cronon and Steven Vogel) the world *is effectively exhausted* by discourse.[13]

The residual metaphysical picture is hostile to nonhuman agency and alterity from the outset because, whilst the epistemological goalposts have shifted, the problematically objectivistic ontological terms extended to nonhuman entities remain significantly intact. They remain intact for the familiar reason that, insofar as it may make substantive claims about the *more-than*-human world, natural science retains little appreciation of the situated fields in which its basic terms are configured and only subsequently reified as transhistorical, acosmic, or "objective." Barad puts this sort of concern as follows: "We do not obtain knowledge by standing outside of the world; we know because 'we' *are* of the world. . . . The separation of epistemology from ontology is a reverberation of a metaphysics that assumes an inherent difference between human and nonhuman, subject and object, mind and body, matter and discourse" (Barad 2007, 147).

None of which is to say that objectivistic ontologies have *no* sphere of justification. Jane Bennett argues that where our survival fundamentally depends upon action-oriented philosophies, there are good practical reasons to temporarily adopt the instrumentalist stance that the subject/object dualism motivates. It is, however, "also dangerous and counterproductive to live this fiction all the time (as Nietzsche and Bergson also note), and neither does it conduce to the formation of a 'greener' sensibility" (Bennett 2010, xiv). Objectivistic schemas fail to motivate a "greener sensibility" because they encourage colonial violence against the more-than-human world. But Bennett's claim also resonates with Evernden's: although an objectivistic stance may have action-oriented merit in *extraordinary* circumstances, it *ordinarily* prevents critical self-reflexivity about how our everyday commitments and attitudes might precipitate our crisis. We might also be reminded of Plumwood's contention that human survival may *rely* on mitigating the consumeristic behaviors licensed by exactly the objectivistic framework that Bennett mentions.

2.2.2 Fact/Value Dualism

Like many ecophenomenologists, most ecofeminists also recognize that the hyperseparation of facts and values is an objectivistic fiction which is implicated in our commodification of the more-than-human world. For ecofeminists, this is primarily because the normative background against which "facts" emerge is structured by dualistic dynamics and metatheoretical assumptions. Ecofeminists thus provide further reason not to uncritically take up the basic epistemological and ontological terms of naturalism, because the natural science which informs it lacks value-neutrality. Plumwood (2002, 43–44) even accuses (what Kuhn might call) "normal science" of being the pinnacle of masculine rationalism because its privileged axiomatic terms present the viewpoint of the occupied center as if acosmic. "Disengagement and neutrality," Plumwood writes, are "as mythological in science as in the market" (2002, 41).

It will be helpful to illustrate Plumwood's point. Beginning in the sixties, an influential movement in the philosophy of science led by Isaac Levi recommended that scientists adopt common "canons of scientific inference" (i.e., "epistemic values" of parsimony, explanatory power, and the like) to determine the appropriate relationships between evidence and hypotheses (Levi 1960, 345; Steel 2010). Levi's cabal (which would later include Kuhn) sought to ensure that science could satisfy the previously unarticulated (but widely accepted) demand to remain free of "nonepistemic" values, despite the inferential limitations arising from its empirical methodology. While Levi's epistemic values may play an important role in good science, however, (chiefly feminist) philosophers of science have observed that their demands can often be satisfied equally well by rival hypotheses employing the same evidence. The Quine-Duhem thesis holds this to be because scientists are historically situated actors whose hypotheses are *always* underdetermined by the available evidence, and so the science cannot *itself* fully arbitrate the plausibility of hypotheses (Duhem 1982; Glazebrook 2005). Merleau-Ponty makes a related point when he contends that "science takes part in history, even nature, and so is not its correlate" (N, 269).

According to this line of argument, the feasibility of hypotheses is determined largely by the scientist's auxiliary assumptions. Donna J. Haraway (1989) argues, for instance, that predominant explanations of primate behavior gain traction through masculine narratives which understand human evolution in terms of the violent taming of the nonhuman environment (which is essentialized and valorized). One of Levi's own canonical pseudo-values is the demand for parsimony, which has historically been pivotal in justifying extensional realist reductions of the more-than-human world. However, as Plumwood argues, it can only do so when married with (objectivistic and dualistic) auxiliary assumptions about the phenomena under discussion, and when other plausible values—like regard for ontological *hetero*geneity—are ignored (Longino 1996, 45–47). The failure of Levi's paradigm supports Plumwood's (2002, 38) argument that (as Heidegger also suggests) to separate technoscience into application (technology=bad) and theory (science=good) doesn't do justice to the epistemic reality. Nor does it foster critical reflexivity about the assumptions and commitments which underwrite the science.[14]

It's important to emphasize, however, that Plumwood's objection to science's alleged value-neutrality goes beyond exposing its falsehood. Plumwood questions the value of a disengaged science, should it even be possible. Following Theresa Brennan, she contends that a "sado-dispassionate" science is problematic insofar as it licenses "ratiogenic domination" of the more-than-human world (2002, 41). The ideal of social and ethical disengagement creates a "commitment vacuum" in science which, she claims, undermines its "ability to resist its cooption by economic forces and works systematically against a science committed to social responsibility" (2002, 41).

We should disambiguate two strands in Plumwood's objection. The first is straightforward: a disengaged science cannot *itself* be socially responsible. Science thus retains an "excess of objectivity" that is "adaptable to partisan politics, rather than enabling wise decision-making" (Glazebrook 2005, 79). A disengaged science can inform us about the volume of crude oil in shale beds, or potential crop yields, but it can say nothing about the axiological merit of pesticide-intensive agriculture or fracking. Even an "ideal" science can, therefore, play only a *regrettably* minimal role (i.e., by providing "dispassionate evidence") in addressing our colonial relationships with the more-than-human world.

Plumwood's second misgiving is more significant: under a sado-dispassionate science (as Levi apparently recognized), allegedly external social and ethical values ultimately dictate the shape of research. At least under dualism, she notes, economic rationality dictates that research which is sufficiently well funded to get under way is determined primarily by instrumental concerns about further research funding, or other economic "goods" (Plumwood 2002, 40). Viable science must ordinarily produce economically useful results, and these results become "the available science." Worse still, only the results of "successful" trials (i.e., research yielding its sponsor's intended result) are routinely published (Douglas 2016, 624). Plumwood cites evidence from Canadian marine science which, until cod fishing collapsed in the nineties due to overexploitation, focused almost exclusively on establishing that "production" was at "safe" levels. Around the same time, Canadian attempts to investigate the impact of overfishing on marine ecosystems were frustrated by insufficient government or corporate funding. Likewise, Hugh Lacey (2005) worries about the unequal resources allocated to developing genetically modified organisms for large-scale agribusiness over the investigation of smaller-scale, lower-impact alternatives. If, under dualism, even a sado-dispassionate science bequeaths us a world skewed toward economic goals, then it is (as Merleau-Ponty and Heidegger claim, for related reasons) further problematic to take even an ideal science to arbitrate the distinction of "real" from "unreal" without reference to the background against which any such reality is inaugurated.

Nevertheless, the "failure" of normal scientific objectivity isn't all bad news. For ecofeminists, Bryan E. Bannon explains, "the divide between ontology and ethics erodes" (2014, 6) in a manner that speaks to our present concern. By resisting a basic commitment to scientific naturalism's extensional realist boundary conditions, ecofeminist theory becomes open to a richer range of phenomena—including axiological ones—*within* the more-than-human world. Independent of a prior commitment to a subject/object dualism, under which each nonhuman entity must be an "intentional nullity, never itself . . . active in disclosing knowledge" (Plumwood 2002, 45), there is no obvious reason to think that such meanings are (fallacious) constructions or projections contained *entirely* within the (human) subject. Ecofeminists are, after all, suspicious of the alleged neutrality of the "raw, 'pure' observational data" (Plumwood 2002, 43) which would permit the scientist to say otherwise. All of this is, of course, theoretically consistent with Merleau-Pontian ecophenomenology. Plumwood, however, highlights a pervasive *androcentric* politics within naturalism that we must be radically reflective about if we're to avoid reinscribing it via any "meanings" we invoke.

Plumwood's overall objection to our dominant naturalistic paradigm targets its demand for epistemological impartiality in the name of "objectivity." For Plumwood, since it is principally dualism's metatheoretical demands which delineate any such "impartiality" or "objectivity," scientific "disengagement" is as politically engaged as anything else. The requirement to return to acts of understanding and knowing as ones with which one is always practically and intersubjectively engaged, and in which one is deeply invested, represents another point of convergence between Merleau-Pontian ecophenomenology and ecofeminism. A bone of contention arises, however, regarding how to reconfigure ontology and epistemology considering this revelation. Plumwood's answer is typical among ecofeminists. She recommends that ontology and epistemology become the preserve of "dialogical," "self-reflective," and "non-reductionist" forms of science (2002, 53). Merleau-Ponty, however, holds that natural science (in any guise) cannot exhaust either ontology or epistemology, because it is problematically limited by objectivistic metaphysical presuppositions at a more fundamental level. It is to this issue that we should now turn.

2.3 PROBLEMS WITH PLUMWOOD'S NATURALISM

2.3.1 Naturalism and Alterity

The problems Plumwood identifies with our dominant philosophical and scientific paradigms result from their ongoing failure to address the hegemonic sediment inherent within their basic (meta)theoretical terms. Insofar as they operate within a naturalized paradigm of dualistic logic, Plumwood and other ecofeminists think, science and philosophy are doomed to reinforce hierarchical and exploitative means of understanding and relating to the more-than-human world. If so, then the ecofeminist identification of a dualistic logic which underpins those terms will further help in tackling our environmental crisis at root. Nevertheless, in mind of our earlier investigation of the relationships between our ontological contentions and phenomenological experience, there remains a tension. Although ecofeminists typically disavow objectivistic *conceptual* schemes, they frequently frame their critical analyses and prescriptions within "suitably modified" naturalistic frameworks.[15] Since, as previously argued, naturalism—as a manifestation of objective thought—is both phenomenologically suspect and hostile to (nonhuman) alterity from the outset, we should have concerns about naturalism's ability to do the work that Plumwood and others ask of it.

To explain: Plumwood is set against dualistic subject/object relations as monological "conceptual frameworks" (2002, 41), and not the basic epistemological distinction between perceiver and world-in-itself which naturalists retain. Which isn't to say that Plumwood chauvinistically denigrates nonhuman forms of agency, for instance, by straightforwardly taking up the for-itself ("culture")/in-itself ("nature") *dualism* as part of her basic metaphysics. For Plumwood, nonhuman entities may be agentic and autonomous in manifold ways (see the following chapter). Moreover, for Plumwood, human subjectivity and agency aren't free-floating or ready-made; both are intrinsically relational in certain epistemic and ontological senses that we've already explored. But there are nonetheless ways that the world simply (one might say, *objectively*) *is*, and which may be conceptually mis*represented*. According to Plumwood, for instance, "nature" just *is* "ecological process," which includes "but greatly exceeds the human" (2001, 29), a truth which the modernist "subject" misrepresents by denying his material relationality. His theoretical contentions fail to correspond with their existent target: nature-in-itself (suitably reconceived). For Plumwood, then, dualistic logic's main threat lies in its capacity to license androcentric *mis*representations of the way the world-in-itself *really* is (albeit in a radically anti-Cartesian sense), which is why Plumwood prescribes what she calls a "progressive naturalism" (2001) to address the commitment vacuum filled by hegemonic ideology when epistemic "disengagement" is promoted in science.

Given that epistemic disengagement is mythical, Plumwood's naturalism is "progressive" in two ways. It first aims to reinstate "emotional," "bodily," and "political" aspects of reality inappropriately excluded under the sado-dispassionate model because of their alleged "subjectivity" (2002, 42). Doing so, Plumwood thinks, requires allowing relations of engagement such as care or respect to function as sound foundations of knowledge relationships. These modes of epistemic engagement may allow previously backgrounded meanings to show *themselves* to be irreducibly real. Plumwood thus offers an "ethico-epistemological proposition . . . that knowledges that do injustice to those who are known do not provide accurate or ethically appropriate forms of knowledge" (2002, 44).

Second, Plumwood prescribes acknowledging "subject-subject" relationships whereby perceiver and perceived become dialogical epistemic interlocutors. Doing so, Plumwood thinks, will open the appropriately engaged scientist to "awe and wonder" (2002, 50) in a manner rendered impossible under mainstream scientific naturalism's instrumentalist stance. The subject/object dualism entails the "closure of the knower to the known, for the knower . . . can change the other to make it conform to desire but . . . cannot be themselves changed by the other" (2002, 42). A progressive naturalism, however, aims to take seriously the thought that more-than-human entities aren't objects to be *possessed* under antecedent human concepts. It takes knowledge constitution to be a more or less reciprocal dialogue which provides the means by which to better address things on their own, rich, terms.

However, we should have concerns about progressive naturalism's ability to achieve this goal. One such concern arises through Merleau-Ponty's observation that our worlds are inaugurated as diacritically interrelated wholes. As Merleau-Ponty's shipwreck neatly illustrates, to alter the diacritics is to alter a whole field's meanings; it is to become "in truth another world" (*PP*, 16–17) rather than a divergent representation of the same extant thing. Plumwood cannot obviously accommodate these insights because, by appealing to the modality of the in-itself, even a progressive naturalism must take most ontologically real properties to be essentially separate from, or antecedent to, perception, even where properties are relational. Where a person *really* has kinship ties with a certain geographical feature (i.e., a nonhuman "subject"), as indigenous peoples often report, for instance, Plumwood still describes these phenomena without any essential reference to the contribution of the (curiously *de*)situated observer in establishing their ontological reality (1993, 182). Plumwood apparently thinks that the world's meanings are dialogically *revealed to* (e.g., human) interlocutors, rather than co*expressed between* body-subjects and world. Thus, Plumwood's naturalism is, as Merleau-Ponty shows, phenomenologically suspect.

But that isn't our major problem. Our major problem comes with the *consequences* of situating perceptual worlds within the modality of the in-itself. Plumwood and Merleau-Ponty agree that acknowledging the irreducible situatedness of perception is essential to any plausible analysis of the ontological, epistemological, or axiological claims derived from it. Both emphasize that this admission speaks against suppressing "the limits and social shaping of knowledge imposed by the knower's identity and their cultural or political 'slant'" (Plumwood 2002, 43). Doing so, for both, means accepting the impossibility of transcendent or transcendental disengagement from the world. For both, knowledge is at least partly constructed, rather than a simple reflective clarification of passively received sensations in the mode of Kuhn's "normal science." Given these shared commitments, an obvious question concerns how to deal with any anthropomorphism which arises from that situated "human conceptual location" (Plumwood 2002, 57). This is where Plumwood's problems really begin.

Plumwood accepts that "any representation of speech-content for a human audience will have to be an interpretation in terms of human concepts, and in that weak sense, a background level of anthropomorphism is always likely to be present" (2002, 57). Plumwood's language is revealing. She evidently takes the world differentially *represented* by heterogeneous entities to be *the same* world-in-itself. This is an implication of the brand of metaphysical realism permitted by her naturalism, without which, taking seriously Uexküll's earlier contentions, it's difficult to make sense of the apparent "weakness" of the anthropomorphism involved. However,

Plumwood herself thereby risks "closing the knower to the known" by extending the basic physiognomic structures of a reality grasped by situated, human, body-subjects to all nonhuman "subjects" as an underpinning constraint upon their representations. At one point, Plumwood even speaks about the intersubjective experience of encountering a snake by a pool as a "conscious sharing and recognition of states of mind," a claim that doesn't sit easily with her contention that "recognition of the other" acts as a "limit on the self and as an independent centre of resistance and opacity" (1993, 157). Where one takes "conceptual location" seriously, what Plumwood seems to think will differ is the specific way in which one cognizes *the* communication the other party offers. The issue is, as it was for Kuhn, about how one conceives *the* world—which retains its own substantive ontological independence from any specific configuration—rather than about how that world is more basically intentionally constituted.

If so, then Plumwood's claim that progressive naturalism rehabilitates naturalism's backgrounding of "bodily" aspects of reality also becomes vulnerable. Plumwood apparently fails to fully appreciate that one's ontological and epistemological claims cannot entirely transcend the intentional grasp generated by one's bodily structures, since they are one's very means of having a world in the first place. The "thing" that shows up in perception, Merleau-Ponty argues, "can never be actually in itself because its articulations are those of our very existence" (*PP*, 146, 320). By attempting to address the modality of the in-itself, Plumwood seems, contrary to her own inclinations, to attempt to transcend her embodiment and, thus, background the intentional contributions her bodily situation makes to *any* such world investigated. The result for progressive naturalism is that, while nonhuman Others might be dialogically engaged, that "dialogue" risks taking place within the framework established by human body-subjects and being conducted on largely humanoid terms

Now, while Plumwood is right that anthropomorphism needn't entail anthropocentrism, neither are these charges entirely unrelated. Where anthropomorphism does violence to nonhuman Others through a basic hostility to their *alterity*, it prevents more positive ways to understand and relate to them. By denying nonhuman Others like the snake the possibility of substantially different intentional realities, Plumwood appears to flatten ontology in a problematically anthropomorphic and, ultimately (at least ontologically), anthropocentric, sense, which delineates the "truths" and "meanings" available in any "dialogue." After all, do Plumwood and the snake really "have a shared basis of understanding and expectation" (1993, 157) *at the level of world-constitution*?

To clarify the root of the problem, it's important to appreciate that Plumwood's call for transspecies subject-subject relationships arises in the context of Sandra Harding's recommendation to turn to disempowered groups to detect androcentric bias in science. But the cases aren't strictly analogous. Women can contest their objectification without experiencing obviously insurmountable barriers to communication about the available phenomena because, as *homo sapiens* sharing significant bodily similarities, women and men will likely share similar enough worlds. However, this isn't obviously true of more dissimilar entities like dung beetles, saplings, or ticks, which plausibly inhabit *radically different* (if not entirely inaccessible) worlds. In a revealing passage, Plumwood recommends that progressive naturalism aim to emulate the subject-subject approach of anthropology, where "highly articulate 'subjects of study' . . . speak of how and under what conditions they would wish to be studied as subjects" (1993, 54). Most nonhuman entities, however, don't obviously possess the communicative rationality to "speak" in this specific sense. If they did—say, to correct our "misrepresentations" of them—especially given objective thought's apparent influence within Plumwood's naturalism, it's not clear we could address their contributions on their own terms.

In response, one might contend that progressive naturalism *does* permit substantially different intentional realities because, through the terms of engagement demanded by subject-subject dialogue, we become open to surprise by how the more-than-human world reveals *itself*. Given that Plumwood's naturalism is motivated by openness to alterity, a tree or tick's intentional, worldly, bonds, for instance, needn't be uncritically derived from the scientist's in the requisite sense to license a charge of anthropomorphic incorporation or assimilation. The trouble with this response, however, is that naturalism's in-principle limitation of intentional realities still appears too stringent. While there may be a degree of difference in how something configures the world, this cannot apply substantively to the stipulative world-in-itself against which any such configurations emerge. We explored this problem in chapter 1, where we saw that, if the lowest level of reduction is the Gestalt, then one cannot situate the Gestalt naturalistically without paradox (see *SB*, 143–44).

Alternatively, one might argue that the charge of anthropomorphism is overdrawn, given the ontological continuity that Plumwood emphasizes between worldly entities. As human, one is also a primate, a vertebrate, and so on, and these commonalities must influence the contours of the ontological reality we inhabit (Plumwood 2002, 133). On Plumwood's account, as I understand it, one is implicated in a dung beetle or sapling's world in the same way that a European is somewhat implicated in the epistemic perspective of an indigenous Amazonian, albeit to a lesser degree. Plumwood argues that there can, therefore, be no hard-and-fast rule about the blanket illegitimacy of cross-species *representation* (2002, 60).

As I suggested in the previous chapter, I think Plumwood *is* right to argue that carnal connaturality facilitates the possibility of engaging with nonhuman Others—in their alterity—to a degree. This thought will garner importance over the next three chapters. Nevertheless, insofar as Plumwood remains wedded to a naturalist purview, any cross-species "representations" must take place *within* the same "world-in-itself" more fundamentally instituted by a particular kind of situated body-subject: here, a suitably progressive ecological scientist. How, as embodied, one coconstructs one's world (and, subsequently, thematizes it) isn't really up for discussion in Plumwood's work, or if it is, it stands in tension with her naturalism. Thus, through her basic commitment to the ontological presuppositions licensed by objective thought, even Plumwood fails to fully appreciate the implications of the fact that theory is irreducibly situated.[16]

Other attempts to rehabilitate progressive naturalism fall into similar traps. According to the version of "hierarchy theory" Warren adopts from ecological science, for instance, naturalism's problems can be mitigated because naturalism is at least compatible with the claim that "there is no ontologically prior . . . ecological description of nature" (2000, 155). Warren thinks that Callicott errs in refusing to permit the ontological irreducibility of entities at different trophic levels and spatio-temporal scales. Different "observation sets" (Warren 2000, 149) call for different apparatuses and yield different phenomena but, since the various phenomena discovered (dynamic ecosystems or subsystems, individual organisms, or holistic energy flows) are strictly *incompossible*, none is more ontologically "fundamental" than another. Warren's approach *intends*, therefore, to take more seriously the situated and partial openings from which ontological and epistemological claims are gleaned. Nonetheless, hierarchy theory remains problematically totalizing. Insofar as it is naturalistic, hierarchy theory still relies upon a singular, "objective" way the ("wholly determinable," "mind-independent") world *is* in-itself, albeit at a certain spatio-temporal scale and trophic level (and, thus, doesn't straightforwardly "admit of only a single correct description"). Like Plumwood, Warren makes little progress on Kuhn's objectivistic model. We need something more radical.

I don't mean to suggest that ecofeminist attempts to address dualism from within naturalism aren't useful. Plumwood certainly helps construct "more generous and helpful" (2002, 54) narratives than are available through reductionist (techno)science. Furthermore, even in anthropomorphizing nonhuman entities, Plumwood provides a politically useful "counter story" (2001, 31) to Cartesian mechanism. In an age of rampant consumerism, these devices are important in stalling our instrumentalist tendencies and turning our attention toward our submerged mass. I contend, however, that Plumwood's commitment to naturalism hampers her attempt to do justice to the thought that "in the context of the ratiogenic complicity of science in over-exploitation, self-critical forms [of knowledge] are surely what we need" (2002, 44). A commitment to naturalism may limit critical self-reflexivity about the intentional objects and sedimented meanings already implicit in one's basic grasp of the world. In fact, by making the world-in-itself a common perceptual object, Plumwood may court some of the colonial sediment she employs progressive naturalism to disrupt. In short: I'm suggesting that, ecofeminism's many virtues notwithstanding, by adopting a naturalist purview, ecofeminists unnecessarily limit the possibility of a praxis of wholly *radical* reflection and the equally radical stance of humility that praxis may motivate.

2.3.2 Dualism as a More-Than-Cognitive Error

To recap: I've suggested that, insofar as ecofeminists are committed to a naturalistic metaphysical framework, they must view dualism as a more or less cognitive error: a misleading conceptual scheme which licenses exploitative mis*representations* of the more-than-human world. Within this naturalistic purview, the scientific paradigm shift allegedly required to disrupt dualism must be understood in a manner similar to Kuhn's. Since the world *itself* apparently doesn't change with a shift in our account of it (although such shifts may, of course, cause material changes), we just need to develop better ways (i.e., via nonexploitative conceptual frameworks) to perceive and theorize the world. That is what progressive naturalism is charged with doing. The problem, however, is that, by reifying the basic ontological entities subsequently accessed by scientists, like Kuhn, Plumwood courts the transcendent or transcendental perspective from which ecofeminists rightly distance themselves.

Ecophenomenologists needn't make this error. Nonetheless, their rejection of any wholly transcendental ontology comes at a cost. For Merleau-Ponty, one's worldly grasp is always infused with the surreptitious sediment of one's situation. That changes how we must think about the sediment that Plumwood foregrounds. In dualistic societies, for instance, Merleau-Ponty would surely admit that dualistic sediment saturates the intersubjective and operative background which generates the meaningful physiognomy of our intentional objects. Merleau-Ponty thereby exposes the true extent of the problem and the challenge in addressing it: dualistic logic doesn't *refer* to the world we experience; it is (at least in "the West") *ontologically inseparable* from that world.

Given Plumwood's attention to "backgrounding," it may seem a strange oversight that she doesn't understand dualism to be an existent aspect of the world itself (rather than our cognitive attitudes toward it). However, according to the metaphysics of presence which objective thought serves to motivate within naturalism, admitting the real existence of something like dualism is tantamount to making it a feature of the world-in-itself and, therefore, (erroneously) naturalizing it. One might follow Merleau-Ponty here in thinking that this issue again emerges out of our failure to recognize the primacy of meaning *co*expressed between chiasmic parties, within a Gestalt-structure.

But why is the metaphysical status of dualistic logic important for our present concern? If, as ecofeminists should surely agree, we habitually grasp the world in somewhat dualistic ways, then it isn't obvious that dualism can be satisfactorily disrupted by cognitive means, as Plumwood implies. The ecophenomenologist Erazim Kohák, for example, argues that attempting to reflectively bracket our instrumentalist prejudices will "prove futile" because "though we bracket the construct of 'nature' as a mechanical system and the human as the sole source of meaning, our urban experience will lead us right back to it" (1984, 23). As body-subjects embedded within an intersubjective lifeworld overlaid with dualistic and objectivistic meanings at an operative level, Kohák thinks, we already relate to that world colonially, as a matter of praxis. And we have probably always done so. Piecemeal moments of reflective clarity, therefore, are unlikely to disrupt our ordinary, colonial, being-in-the-world. A related concern leads Linda Martín Alcoff to ask, rhetorically, "if identity prejudice operates via a collective imaginary, as [Miranda Fricker] suggests, through associated images and relatively unconscious connotations, can a successful antidote operate entirely as a conscious practice? Will volitional reflexivity, in other words, be sufficient to counteract a non-volitional prejudice?" (2010, 132).

Recent research on implicit bias substantiates Alcoff's and Kohák's concerns. One of its key findings is that "simply trying very hard to be unbiased . . . will do nothing to combat either implicit bias or stereotype threat. In fact, research shows that this may *increase* implicit bias" (Stewart and Payne, cited in Saul 2013, 50). Given Plumwood's skepticism about "Western" logocentrism's ability to reconfigure rationality, this revelation isn't entirely surprising. Nevertheless, it exposes the limitations of Plumwood's naturalistic purview for bringing about radical change, a limitation which Plumwood gets close to admitting when she says that "rationalist influence makes philosophers treat centrism as if it were a matter of cognition. . . . Centrism is tested by behaviour rather than avowal" (2002, 98–99). Plumwood, nevertheless, still advocates "studying up" (i.e., making -centric logic conspicuous to theory) as *the* method by which to address dualism. We shall explore alternative means in chapter 3, but it will suffice to say for now, with Merleau-Ponty, that if dualistic or objectivistic assumptions are ingrained, more-than-cognitive errors, then our means of addressing them must also be more-than-cognitive, because substantive change requires a more fundamental shift in praxis.

By retaining the problematic theoretical kernels identified above, Plumwood and other ecofeminists are prevented from entirely avoiding "a new narrative of dominance that mimics and parallels the old, failing to imagine with sufficient daring the identities and relationships the original narrative created" (2001, 8). However, I see no *in-principle* reason why Plumwood's insights cannot be reconfigured ecophenomenologically.[17]

2.4 CONCLUSIONS

Ecofeminists helpfully foreground some hegemonic sediment which may imbue our problematically objectivistic ways of encountering the more-than-human world. Moreover, since the dualistic logic seemingly enshrined within our default scientific naturalism is inherently exploitative, Plumwood, in particular, helps clarify why its terms are so problematic for our ongoing worldly relationships. By emphasizing that any analysis of environmental "issues" and proposed "solutions" must address their underpinning colonial logic, ecophenomenologists and ecofeminists are also united in their determination to investigate the assumptions and commitments constitutive of our submerged mass. However, insofar as they are constrained by more fundamental objectivistic metaphysical stipulations, many ecofeminists appear unduly limited in their capacity to do so. Nevertheless, given their significant intertwinings, ecofeminist and ecophenomenological approaches appear to be prima facie reconcilable in a manner that favors reconfiguring the former's insights into the latter's praxis. In the next chapter, we shall explore how such a praxis might disrupt the objectivistic underpinnings of our environmental crisis.

CHAPTER THREE

SEEING BETTER

I've hitherto argued that certain objectivistic commitments underwrite the violent behaviors responsible for our crisis situation. Merleau-Ponty provides reason to believe that the objectivistic metaphysical assumptions of our default naturalism limit our capacity to appreciate the more-than-human world on its own, rich, terms. As a manifestation of objective thought, naturalism requires or presupposes transcendent or transcendental access to the modality of the in-itself. Naturalism, therefore, lacks the capacity for sufficient critical self-reflexivity about the contribution of one's situated and embodied grasp to the world's intentional physiognomy. Given its narrow metaphysical scope, scientific naturalism is also problematically limited in the reductive terms it can extend to the more-than-human world's "objective" furniture. We've seen ecofeminists like Val Plumwood further problematize the terms scientists ordinarily extend to the more-than-human world because they are underpinned by a colonial logic which sets humans (qua subject) and nonhumans (qua object) apart in kind. Thus, according to the line of argument I've been pursuing, not only are more-than-human "objects" misconstrued as metaphysically distinct entities, the reductive terms imposed upon them ensure their continued exploitation.

I've also suggested that tackling the imposition of onto-epistemological violence requires a *radical* sort of reflection that is best facilitated by rethinking our environmental crisis ecophenomenologically. Merleau-Pontian ecophenomenology is particularly promising for our task because it foregrounds the embodied habituality of perception, without requiring that perceptual habits be mechanistic or deterministic. Thus, as I suggested in chapter 1, the radically reflective Merleau-Pontian body-subject might literally *learn* to perceive better. For Merleau-Ponty, one perceives *better* where one more accurately expresses the more-than-human world's meanings, without requiring that expression reflectively coincide with any (problematically objectivistic) world-in-itself. Since, on Merleau-Ponty's account, new habits may be sedimented into one's basic worldly grasp, Merleau-Pontian ecophenomenology offers promise that one might come to perceive *and* relate to the world in more positive (and less objectivistic) ways as a matter of praxis.

In this chapter, I explore how Merleau-Pontian ecophenomenology might provide resources to facilitate the above task. It is split into three sections. The first explores how such an ecophenomenology informed by ecofeminist (and new materialist) insights might foreground subject-object *continuities* as correctives to scientific naturalism's insidious dualism. Doing so will help us disrupt tendencies toward possessing and controlling the more-than-human world—in both conceptual and behavioral senses—ordinarily legitimated by a subject/object dualism. The second section explores how such insights speak in favor of adopting a new, nonreductive, onto-epistemological norm. The details will come later, but my basic argument is as follows: if the naturalistic supposition of ontologically primitive "subject" and "object" poles rests on erroneous (meta)theoretical assumptions, then we should reject any correspondence theory of truth which relies upon them. Since the world may, plausibly, be taken up under heterogeneous schemes of description without individually exhausting the meanings available for expression, then we should replace the naturalist's totalizing "God's Eye view" with the "*view-from-everywhere*." That is, the *norm* defined by the sum of all meaningful expressions of the world, *including nonhuman ones*. In the final section, I entertain some suggestions about how one might inculcate habits that facilitate the sustained possibility of seeing *better*. I explore the roles of attention and, particularly, hesitancy—conceived as affective openness to the indeterminacy of perceptual habits—in entrenching a radically reflexive praxis. I ultimately argue that the indeterminacy of perceptual habits provides reason to be optimistic about our prospects for doing so.

3.1 RETHINKING SUBJECT AND OBJECT

I've argued that we must disrupt our objectivistic assumptions and commitments if we're to reconfigure human-world relationships outside the rubric of the latter's domination. Plumwood recommends we begin by addressing the exclusionary logic which legitimates hyperseparating "subjects" and "objects" from the outset. A "non-reductive resolution," she contends, "requires both that we reconceive ourselves as more animal and embodied, more 'natural,' and that we reconceive nature as more mindlike than in the Cartesian conception" (1993, 124). Nonetheless, I've suggested that any attempt to rethink "subject" and "object" terms with sufficient daring cannot retain a naturalistic purview without threatening nonhuman alterity via one's imposition of the terms of a "world-in-itself." Uncovering subject-object continuities must, therefore, involve more than an uncritical appeal to the reality enframed by progressive scientists. Nevertheless, this admission needn't make it impossible for ecophenomenologists to address the more-than-human world on its own terms. Merleau-Pontian ecophenomenologists outright deny the subject/object bifurcation which might otherwise foreclose the possibility of access to it. In this section, then, I explore some possibilities for subject-object continuity which emerge by engaging with the meaningful terms under which the more-than-human world makes *itself* available for expression by the radically reflective ecophenomenologist.

Continuities may be recognized in a "top-down" way, where one addresses "subjective" or mindlike properties in more-than-human "objects." Under a "bottom-up" strategy, conversely, one addresses the materiality of mind or subjectivity. Beginning with the former, Plumwood argues that important continuities may be unearthed by *taking care* to register the agency, teleology, and intentionality ordinarily denied to material entities. Intentionality and agency are especially difficult to identify under scientific naturalism, where nonagentic efficient causation is typically considered the only natural (rather than implausibly supernatural) power. However, we've seen that natural science's underlying assumptions and apparatuses reflectively distort or unfairly exclude some of the meaningful phenomena available for expression. Identifying where mindlike traits might be attributed to nonhumans will assist our groundwork in two ways: first, by disrupting the bifurcation which licenses a subject/object, or mind/nature *dualism*; second, by problematizing the totalizing assumption that the "natural" world consists entirely of determinate "objects," which may be epistemologically *possessed* and controlled in the requisite manner to license their habitual instrumentalization.

3.1.1 More-Than-Human Intentionality

Let's begin with intentionality. Both Merleau-Ponty and Plumwood (1993, 131) follow Franz Brentano in understanding intentionality to be irreducible to representational consciousness. Unlike Brentano, however, they do so partly to avoid unfairly precluding entities like ticks from the sphere of mindedness. For Plumwood (like Brentano), intentionality refers to myriad criteria like goal-directedness which cannot be accommodated mechanistically. We've also seen Merleau-Ponty exemplify dung beetle and feline intentionality in this manner. In these cases, Merleau-Ponty contends, "we no longer see where behavior begins and mind ends" (*N*, 178). Although Merleau-Ponty argues that more complex forms of intentionality aim less at goals and more toward the interpretation of symbols (between language users, for instance), this observation cannot support a differentiation in kind. "There is not," Merleau-Ponty argues, "a break between the planned animal, the animal that plans, and the animal without plan" (*N*, 176). As we saw in chapter 1, behaviors throughout the intentional continuum cannot be adequately explained by causal imposition of the way the world currently, "objectively," *is*, even apparently at the minimal cellular level.

Their transspecies attributions of intentionality shouldn't, however, be confused with a romantic, anthropomorphic, projectionism divorced from rigorous scientific study. As Louise Westling (2014, 83) notes, Merleau-Ponty's account chimes with that of contemporary microbiologists. According to "niche construction" theorists like Richard Lewontin (2000) and John Odling-Smee (2009), for instance, even relatively simple organisms configure, select, and adapt salient features of the environment into which they ambiguously interlace and coconstruct. Earthworms, for example, adapt soil to their water-intensive physiological demands by "tunnelling, exuding mucus, and eliminating calcite" (Odling-Smee 2009, 74). This process also provides structural and nutritional benefits for plant growth and, subsequently, the worms themselves.

However, although they hold organism-environment dialectics to be mutually transformative, these microbiologists remain beholden to the restrictive metaphysical assumptions of the naturalistic frameworks they employ. Odling-Smee, for instance, rightly criticizes the "Newtonian residue" which requires standard evolutionary theory to understand organisms as "passive objects," causally determined by determinate external environments (2009, 77). Nevertheless, Odling-Smee still attributes any "non-random work" in earthworm niche-construction to the mechanical effects of "naturally selected information carried in their genes" and the determinate physicochemical properties of "objective" stimuli on their chemoreceptors (2009, 77). Merleau-Pontian ecophenomenologists, however, may better address naturalism's "Newtonian residue" by understanding the earthworm to intentionally engage with the phenomena "motivated" (rather than "objectively" efficiently caused) by the meanings specific to the Gestalt it coconfigures (*PP*, 31, 50). By eschewing naturalism's reductive presuppositions,

Merleau-Pontian ecophenomenologists may thereby afford earthworms the irreducible intentional teleology they appear to possess, without extending them an implausible reflective consciousness.

Although Merleau-Ponty focuses on animal intentionality, he's amenable to extending intentionality further (*SB*, 123).[1] Likewise, Plumwood finds intentionality exhibited by individual trees, which "appear as self-directing beings with an overall good or interest and a capacity for individual choice in response to their conditions of life," or "forest ecosystems [which] can be seen as wholes whose interrelationship of parts can only be understood in terms of stabilising and organising principles, which must again be understood teleologically" (1993, 135–36). Plumwood's contentions too are well supported by recent plant science. Examples here are more surprising, ranging from "kin recognition" behaviors between plants to "communication networks as extensive as whole forests in size," through which information about predators is shared and preemptive responses are prepared (Trewavas and Baluška 2011, 1225).

The research cited above has, nonetheless, proven controversial among plant scientists. The plant physiologist Lincoln Taiz, for instance, accuses so-called plant neurobiologists like František Baluška of "teleology, anthropomorphizing, [and] philosophizing" (Taiz, quoted in Pollan 2013). However, each of these "charges" is revealing insofar as they speak to our concerns about naturalism's limited critical self-reflexivity. In the previous chapter, for instance, we problematized a natural science detached from all "philosophical" commitments, especially when that science is couched in the sado-dispassionate and positivistic terms that Taiz thinks uniquely suitable. Likewise, the charges of "teleology" and "anthropomorphizing" appear to assume an equally problematic in-principle distinction between nonhuman "objects" and humanoid "subjects." But Taiz's problems run deeper than this. His main objection to Baluška's attribution of plant intentionality revolves exclusively around apparently substantial differences between "electrical signalling" in plants and the "true neural mechanisms" of "higher" animals (Taiz, quoted in Pollan 2013). Like many of his "Alpi letter" colleagues, Taiz doesn't appear to be reflexive about the objectivistic (meta)theoretical assumptions which underwrite his more fundamental commitment to mechanism.[2] Hence, for reasons previously explored, his objections do little to undermine the plausibility of my present contentions.

In any case, while the attribution of nonhuman intentionality or mindedness remains controversial, it coheres with other murky boundaries between stipulative "natural kinds" that many natural scientists already concede. These include the fuzzy living/nonliving binary (which even Richard Dawkins acknowledges), and the animal/plant binary undermined by photosynthesizing slugs (Morton 2010, 276; Angelin 2015, 346). These continuities might help upset charges of "anthropomorphism" or "teleology" which rely upon hard-and-fast distinctions. However, resonances between the implausibility of a straightforward subject/object binary and scientists' wider taxonomic concessions seem to have been conveniently backgrounded for reasons we explored with Plumwood in the previous chapter.

Nonetheless, emphasizing nonhuman intentionality is important. Most obviously, nonhuman intentionality problematizes the subject/object dualism which licenses anthropogenic colonialism. Plumwood has argued, recall, that the dualistic restriction of mindedness or subjectivity pacifies nonhuman entities and thereby erodes their capacity to resist objectification and commodification. Doing so, however, requires radically excluding a separable order of "objects" which more-than-human intentionality problematizes. Simply put: even earthworms and forests don't obviously differ in kind from human "subjects" in the definitive sense to license their denigration, even under neo-Cartesian mechanisms like the Differential Imperative.

Perhaps more important, sensitivity to nonhuman intentionality may disrupt colonialism because acknowledging other intentional realities—ontologically real meanings irreducible to one's own—undercuts the very possibility of a totalizing onto-epistemological stance. Plumwood (2002, 177) gives the example of a forest ecosystem whose irreducibly intentional goods are incompossible with the economistic or technoscientific schemas proffered by humans. The latter not only construe the forest wholly in terms of their hegemonic interests (as a capital, fuel, or oxygen resource), their restrictive apparatuses also preclude *narrative goods* which, for trees and earthworms, appear to be irreducibly real, yet reach into the future. Furthermore, since these narrative goods aren't obviously discernable as things-in-themselves, they expose the limitations of taking positivistic apparatuses to exhaust reality and, thus, arbitrate *the* terms under which environmental "issues" are understood. This is because the physiognomy of intentional objects ultimately depends upon their situated and exclusionary terms of engagement.

Moreover, Merleau-Ponty has argued that heterogeneous intentional realities are different worlds, rather than divergent interpretations of the same determinate content. They cannot, therefore, be exhaustively enframed by even an ideal science without significant distortion. Jakob von Uexküll's ticks and Merleau-Ponty's canines neatly demonstrate that the world's basic configuration is intimately related to the *embodied* grasp that one has upon it. Given our limited and partial openings onto the world, some of the meanings available there may differ radically from our own and some, as our ticks and canines also illustrate, must simply be beyond our ken. Nonhuman intentionality, in short, speaks against the project of mastery crystallized in the scientific naturalist's appeal to a purified realm of determinate things-in-themselves. The naturalist's objectivity of the God's Eye view can only really be sustained by violently subsuming reality under a peculiarly anthropocentric—and, if Plumwood is right, *androcentric*—conceptual scheme.

3.1.2 Panpsychist Problems

There may, however, be problems with the "weak panpsychism" Plumwood takes the widespread distribution of intentionality to imply (1993, 133; 2002, 54). To explain: one reason for our overall objection to scientific naturalism is its failure to balance "the recognition of continuity with the recognition of difference" (Plumwood 1993, 125). Naturalism typically neglects important ontological continuities in carnal embeddedness or material relationality between the (human) "subjects" and material "objects" it hyperseparates. Mainstream scientific naturalism is also hostile to difference because it subsumes material "objects" within a certain objectivistic purview (ordinarily, that of extensional realism), and reifies the terms of epistemic engagement specific to that purview as if transhistorical and universal (albeit, perhaps, with the usual Kantian caveats). Our attention to nonhuman intentionality is, therefore, motivated primarily by the attempt to juggle these requirements sufficiently well to engage with more-than-human entities *in their alterity.*

By extending intentionality *universally,* however, we risk inappropriately overplaying continuity. Plumwood emphasizes that the panpsychist attribution of intentionality cannot unproblematically entail the mere extension of properties of human consciousness. Any such "strong panpsychism" assumes that "mind is either totally present in a human or humanoid form or totally absent," which Plumwood rightly rejects (1993, 133; 2002, 189). A related false dilemma lurks behind Taiz's charges. Nevertheless, as Merleau-Ponty's work illustrates, we grasp our own intentionalities as reflective consciousnesses operating within certain kinds of Gestalt structure. Ascriptions of nonhuman intentionality, therefore, *risk* being derivative and misleading extensions of humanoid consciousness, especially when, as in Plumwood's panpsychism, one's theoretical commitments *require* its universal application to divergent entities like neutrinos, marine ecosystems, or *E. coli* bacteria.

Given the likely implication of objectivistic sediment in the violence of our crisis situation, there are good political reasons for generally beginning from an intentional stance, rather than one organized to exclude nonhumans from the intentional sphere. However, we should *take care* not to push this too far. Plumwood does so when her panpsychism leads her to ascribe the same sort of teleology or narrative desire to whole mountains, stones, and glaciers that she ascribes to snakes.[3] Plumwood claims, for instance, that damming a glaciated valley is problematic because it prevents it from "telling its story" (1993, 138). Attempting to engage with such entities on nonreductive terms is important for environmental theory. Truly expressing the world is, after all, what radical reflection aims at. But by universally imposing atomic intentionality, one risks a theory-guiding, anthropomorphic projectivism that connotes a push toward the very onto-epistemological imperialism that Plumwood sought to mitigate *by* engaging with nonhuman mindedness. This is because, if nonhuman entities are recovered only on terms which largely reflect our own, then we'll likely read our own meanings, intentions, and desires into them.

We previously saw glimpses of *something like* a tacit imperialism through Plumwood's claims about cross-species translation, resulting from her more fundamental commitment to naturalism. There, Plumwood's interpretations of the snake's dialogical interjections implied a problematic underlying anthropomorphism because Plumwood was insufficiently *attentive* to the plausible extent of intentional heterogeneity. Where one "considers" an absent intentionality, a more pronounced ventriloquism will likely be effected, especially given the extent to which dualistic logic infuses *our very worlds* as we grasp them from situated master perspectives. None of this sits easily with Plumwood's claim that through weak panpsychism's "intentional stance we open ourselves to possibilities and exchanges which are not just of our own devising" (1993, 137).

Plumwood (2002, 58) seems to be wrong, then, to imply that the anthropomorphic attribution of glacial intentionality, for instance, is at least not a dangerous error. The intentional "goods" of the glacier, marine ecosystem, snake, or rainforest—*as one understands them*—will at some point be weighed against others where not all can be practically pursued. Plumwood *defines* "ethics" as "the domain of response to the [more-than-human] other's needs, ends, direction or meaning" (1993, 138). However, if panpsychism is hostile to nonhuman alterity in the manner I contend, then what panpsychists take to be a dialogical solution to an ethical quandary may mask a colonial monologue. This point supports a heightened focus on radical reflexivity rather than a blanket, albeit speculative, panpsychism.

But, as we saw in the previous chapter, this sort of problem isn't limited to panpsychism. It pervades the wider ecofeminist literature which employs dialogical accounts of the human-nonhuman relationship that fail to sufficiently problematize the terms under which nonhuman interlocutors (or their alleged intentions) are understood. One example is Karen J. Warren's (2000, 102–3) account of rock climbing in a manner which allegedly *exemplifies* the reciprocal care and respect characteristic of an anticolonial environmental ethic. However, Molly Hadley Jensen observes, Warren never "limit[s] or alter[s] her activity; she continues to climb. One could deduce from her account, that an ethic of friendship involves only a new sentiment toward non-human nature which may assuage human feelings of guilt but leave patterns of use and abuse undisturbed" (2002, 42). Stacy Alaimo diagnoses a similar problem in Carolyn Merchant's account of *nonhuman* nature as a "free, autonomous actor," a characterization Merchant uses to prescribe a "partnership ethic" which relies on hyperbolized accounts of humanoid autonomy, and which cannot, therefore, "flourish within models of interdependency, ecological systems, or environmental health" (Alaimo 2007, 246; Merchant 1996, 221). Theirs is a particularly dangerous kind of error which may *discourage* genuinely radical reflection. Like Plumwood, Warren and Merchant seemingly try "to make the other's unfamiliarity familiar, and in doing so become blinded to the ways these perceptions conflict with the very misunderstandings about them that, ironically, one seeks to address by imaginatively representing them" (Johnson 2008, 175).

An alternative interpretation of Plumwood's panpsychism helps little. Bryan E. Bannon holds that, for Plumwood, the intentionality in question is really *more-than*-human nature's (conceived holistically as an autonomous, teleological, center of life) rather than belonging to the glacier as such. Nonetheless, Bannon argues, Plumwood still thereby "personif[ies] nature as an agent with projects rather than confronting it in distinctively nonhuman terms" (2009, 48–49). By failing to problematize the hierarchical assumption that the "nonmental" must be a privation, Plumwood threatens her own act of incorporation. Moreover, although the atomic intentionality that Plumwood employs may be necessary to attribute mindedness to nature-in-*itself*, the reflectively purified account of "nature" thus retained sustains modernism's antirelational assumptions about identity, autonomy, and causality. Thus, not only does Plumwood risk reintroducing the terms of the subject/object dualism via a truncated (subject-subject) reversal, like Merchant, she also apparently wants to (somewhat inconsistently) use its terms to "isolate nature from its relations to humanity," in at least an epistemic sense, "such that it can serve as a normative foundation for a variety of social projects" (Bannon 2014, 35).

A third interpretation, however, seems closer to the spirit, if not always the word, of Plumwood's project. This interpretation takes Plumwood's initial characterization of her panpsychism as paradigmatic: "the view that *mind-like* qualities are to be found in nature, that there is no basis for an absolute break or an unbridgeable gulf marking humans as mindhavers off from the rest of the universe" (1993, 133). Although it operates under a misnomer, *this* weak panpsychism contends that the heterogeneity of mindlike qualities variously enjoyed entails the impossibility of a sharp Brentanian distinction; it resists globally adjudicating which entities are minded (intentionally or otherwise).[4] Such totalizing claims become implausible once that heterogeneity is considered in any depth. Under this interpretation, Plumwood wouldn't be pressed into a problematic anthropomorphism by the apparent requirement to universalize (humanoid) intentionality, nor need she necessarily purify nonhuman entities of their epistemic relations with humans. Plumwood's account would also, therefore, become entirely compatible with Merleau-Ponty's and, as we shall see, Jane Bennett's.

3.1.3 Quasi Agency

A suitably radical rethinking of the subject-object dualism cannot, however, rest with nonhuman intentionality, since further (more ambiguously) "mind-like" qualities permeate the more-than-human world than either Plumwood or Merleau-Ponty acknowledges in any depth. In the present philosophical climate, the task of identifying them has fallen largely to "new materialists" like Manuel DeLanda, Karen Barad, and Jane Bennett, whom I will focus upon here.

Bennett's recent research is guided by the question: "How would political responses to public problems change were we to take seriously the vitality of (nonhuman) bodies?" (2010, viii). Her major work on the subject, *Vibrant Matter,* addresses "active powers issuing from nonsubjects" (2010, ix). By investigating the impersonal agency of entities like dead animals and processed foods that cannot be even ideally reified as "subjects," Bennett aims to disrupt the objectivistic presupposition that politics is a solely human realm. There are, she thinks, important political agents in our everyday experience which aren't "persons" under anyone's estimation.[5] Even "nonsubjects" are, therefore, irreducible to objects, at least as "objects" are ordinarily enframed.

Bennett's claim isn't that nonsubjects configure their own inassimilable intentional meanings. Her political contentions are linked to the reasons she provides to think that scientific naturalism's subject/object binary is "badly made": first, because agency isn't a straightforwardly "subjective" or "supernatural" property in the requisite sense to be satisfactorily purged from a purified "objective" realm; second, because even nonsubjective material entities may resist their objectification under any *totalizing* schema, since "thing power" (2010, 4) is neither mechanistic nor ideally predictable in the requisite manner to permit its ontological or epistemological possession.

Bennett argues, for instance, that paying attention to piles of rubbish can disrupt our usual metaphysical hubris by making us recognize that these "things" cannot be satisfactorily *reduced to* a determinate collection of physico-chemical properties or their utilitarian roles as intentional objects within human narratives. These things may call to us in their differential quasi-agentic capacities as *that* plastic bottle or rat corpse "existents in excess of their association with human meanings, habits, or projects" which have the power, in conjunction with other "actants," to produce unpredictable effects (2010, 4). The term "actant" is Bruno Latour's. In Bennett's example, it may refer to the mode of sunlight, or distribution of other bodies (e.g., bottle cap or glove) as well as the attention of the perceiving body-subject. The unpredictable effect Bennett herself experienced was the disruptive revelation of material quasi agency, but similar actants may also call the perceiver to be struck by the callous wastefulness that collection betrays, and so on. Echoing Merleau-Ponty, Bennett argues that hers is only one way that these "things" may coexpress *themselves*, should one slacken one's intentional threads sufficiently to notice. We must nonetheless emphasize that Bennett's specific example is important in illustrating that nonsubjective quasi agency may reveal itself *phenomenologically*.

But the extent of nonsubjective "thing power" may extend well beyond the capacity to make us perceive differently. "Spontaneous structural generation" (DeLanda 2000, 16), for instance, occurs when chemical systems in far-from-equilibrium states "opt," in a nonlinear manner, for one developmental path or another. This capacity for spontaneous self-organization also imbues cases of "nonlinear combinatorics" (2000, 17), where inorganic compounds acquire novel agentic powers which cannot be further dissected without misrepresentation. DeLanda's (1995) paradigm example of the latter is how the dynamic behaviors of populations of imperfections in crystalline structures have the emergent ability to determine metallic brittleness. Since these microlevel behaviors are strictly unpredictable, contemporary swordsmiths focus not on controlling them, but on facilitating or restricting their movement, depending on the properties they seek. Bennett herself documents "not-fully-predictable encounters between multiple kinds of actants" through the nonlinear behavior of biota like slime mold colonies and the more heterogeneous collectives involved in the coalescence of mineral bodies in evolutionary processes, or the intra-agency of power grids (2010, 10–11, 28, 97, 100).

It's important to emphasize, however, that by appealing to quasi-agentic unpredictability, unlike Merchant (2003), Bennett isn't simply advocating a shift from mechanism to chaos or complexity theories. These sciences still aim at discovering "the law governing irregularities in nature, the hidden structures in apparently random systems, and underlying deterministic causes" (Cook 2011, 149). As Deborah Cook notes, given ecofeminism's imperative to resist hubristic attempts to dominate or objectify "nature," Merchant's enthusiasm about their ability to disrupt our default metaphysical androcentrism is misplaced. Bennett, however, is sensitive to the tendency toward mastery manifested through the totalizing stance these sciences retain. This caution should be read into her appeals to DeLanda's work in nonequilibrium thermodynamics, for example.

Instead, according to Bennett, "bodies enhance their power *in* or *as a heterogeneous assemblage*" (2010, 23).[6] For Bennett, quasi-agentic power, augmented through its distribution across assemblages, is the source of genuine agentic novelty. An assemblage's components may contribute to its emergent powers to effect novel change but, taken discretely, they cannot *themselves* individually or collectively constitute its efficient cause. Bennett explains how understanding a large-scale power blackout in the United States, for example, requires reference to the "distributive agency" (2010, 31) of the whole assemblage. This assemblage includes coal, steel, neoliberal dogma, energy legislation, and electricity (which, insofar as electron movement isn't entirely linear or predictable, itself displays a more discrete quasi agency). The (quasi)agentic capacities of each of these things problematize their objectification because they too retain an *excess* that prevents thematic coincidence with the terms of a totalizing science. As Bennett puts it after repeating Theodor W. Adorno's dictum that "'objects do not go into their concepts without leaving a remainder' . . . life will always exceed our knowledge and control" (Adorno 1973, 5; Bennett 2010, 14).[7]

To further explain, we must emphasize that, for both DeLanda and Bennett, quasi agency is *distributive*. Quasi agency is a kind of "field" not straightforwardly owned by any actant. In Bennett's rubbish example, for instance, her objectifying gaze was disrupted by a confederacy of entities. More pertinently, given the context, Bennett also argues that, within the hyperconsumeristic paradigm of "American materialism," we're so surrounded by "disposable" (and disposed) bodies that *these* assemblages ordinarily produce "*anti*material" effects; they conceal "the vitality of matter" (2010, 5), thereby licensing our uncritical objectification of it. But even these assemblages are mosaics of ontologically heterogeneous bodies, across which (quasi) agency is distributed. Thus, for Bennett, thing power is essentially *relational*.

This is important. By drawing our *attention* toward the relationality of (quasi) agency rather than focusing exclusively upon intentional individuals, Bennett may address the residual atomism which punctuates Plumwood's account.[8] This shift of focus also facilitates gains in recognizing continuities *and* differences between more-than-human entities. It may do so by acknowledging that even nonsubjects like gloves and concrete may escape our thematizing grasp, without their resistance requiring the attribution of a narrow intentional stance. Bennett's investigation of "mindlike" qualities isn't, therefore, vulnerable to the neo-Cartesianism that uncharitable readers might locate in Plumwood's work.

Bennett also corrects Plumwood's and Merleau-Ponty's worrying tendency to treat *inorganic* entities as mere use-objects. One of Plumwood's telling examples is of a Coke bottle which, lacking its *own* intentional context, allegedly "has no direction of travel, and all outcomes with respect to it are indifferent" (1993, 138) until it features within another intentional narrative. For Bennett, however, even discarded plastic bottles retain *their own* quasi agency. We saw as much via the bottle's role in Bennett's revelation about the rubbish. More politically salient manifestations of its thing power might include its capacity to leach poisons into groundwater supplies, or its capacity, when burnt as waste, to become part of a carcinogenic assemblage which may literally become part of one's flesh, or pass (with disastrous consequences) between bodies through breastfeeding (Tuana 2007).

To limit the bottle's political importance to its contingent situation within an intentional horizon carries a faint whiff of residual dualism, especially since the issues outlined above most often befoul dualism's underside. Poorer people are, after all, disproportionately likely to work in plastic factories or to live in areas directly affected by mass waste management strategies (Tuana 2007, 203). The larger-scale effects of these mechanized industries—anthropogenic desertification, flooding, and so on—are also suffered particularly keenly by "non-Western" people, not to mention the gazelle or addax which face extinction. If even nonsubjects cannot be reduced to their "determinate" physico-chemical properties, to their utilitarian roles in others' intentional narratives, or to social constructions without *novel* causal powers of their own, then recognition of their quasi agency will play an important role in any suitably radical reconfiguration of our naturalistic purview.

Nonetheless, we shouldn't abandon our previous insights and take up Bennett's actor-network theory wholesale. I explore some concerns about Bennett's naturalism in section 3.1.9, but here is another: while Bennett is surely correct that nonsubjects might have goods that any ethically and politically sophisticated praxis should register, she cannot obviously permit them *narrative* goods like those required to explain the teleological orientation of (quasi) subjects like earthworms and forests. To account for their behaviors, we require recourse to the kind of nonhuman intentionality which Bennett distances herself from. Although, like DeLanda, Bennett takes actor-network theory to be compatible with intentionality, the only references to intentionality in *Vibrant Matter* concern its cautious attribution (in a *significantly* reduced form) to humans (2010, 32). However, as previously argued, by eschewing or dramatically limiting nonhuman intentionality, we sacrifice something important in our radically reflective account of the more-than-human world. Thus, in taking better stock of it, we need to heed intentionality *and* other "mindlike" qualities.

3.1.4 An Aside: The Anthropocene

Although it isn't our main focus, I'd like to pause momentarily to consider how the preceding analyses might expose something troubling about contemporary Anthropocene rhetoric. The literature is vast and arises from a range of fields and concerns, but I want to focus on one specifically philosophical concern: namely, whether acknowledging the "epochal" shift crystallized under the Anthropocene moniker affords better means of foregrounding and responding to the profound effects that human beings have upon the more-than-human world.

Earlier, I suggested that the evidence cited by Anthropocene proponents (ecosystem collapse, mass species extinction, and so on) exposes our deeply problematic relationships with the rest of the world. However, insofar as it makes claims about "epochal" significance, pro-Anthropocene rhetoric often goes beyond acknowledging the massive extent of human influence and draws more sweeping *metaphysical* conclusions. Following Bill McKibben's (1989) famous contention that anthropogenic changes to the Earth's chemistry signal "the end of nature," the ecologist Erle Ellis, for instance, claims that "from a philosophical point of view, nature is now human nature; there is no more wild nature to be found, just ecosystems in different states of human interaction" (2011, 1027). Likewise, we've seen Paul Crutzen refer to a "*formerly* natural world." Taking it beyond its original geological context, many such proponents seem to think that the definitive reason to recognize the "Anthropocene" epoch lies in its acknowledgment of the anthropogenic *termination* of nonhuman nature.

To understand what's problematic about this claim, we must first ward off a straw man. My point isn't so crude as to highlight the massive anthropocentric hubris involved in claiming that nonhuman nature has been definitively "ended" in the sense of being ontologically exterminated or left causally inert. As Simon Hailwood (2016) clarifies, nobody involved in the Anthropocene debate thinks that human contributions *alone* increase greenhouse gas emissions, for instance. Cattle and rice plants contribute, as do microorganisms that don't significantly populate human bodies. The aforementioned Anthropocene proponents claim that human behavior has—via intensive agriculture, urbanization, and so on—shaped these contributions sufficiently to cease speaking about nature in a "nonhuman" sense.

We should have serious reservations about *this* sort of claim, which I will demonstrate by focusing on one recent (and superficially plausible) version of it. The political theorist Manuel Arias-Maldonado holds that the Anthropocene epoch's allegedly characteristic "socionatural hybridization" is compatible with the "end of nature" hypothesis because "nature is not autonomous from [human] society anymore" (2016, 68–69). Like Steven Vogel (2002, 33), Arias-Maldonado thinks that the impact of anthropogenic processes, from the domestication of wolves to rising sea levels, shows we've used human agency to, literally, construct "nature" wholesale, albeit sometimes in unforeseen ways. For Arias-Maldonado, we rightly speak of nature's "construction" because, while the Anthropocene doesn't herald the end of nonhuman causal powers, within the remaining human-nature assemblage, "human agency remains the main driver for the transformation of nature" (2016, 68). Thus,

Arias-Maldonado thinks, it is appropriate, liberating, and politically useful to adopt the Anthropocene term to denote a *postnatural* epoch.

However, Arias-Maldonado thereby invokes a troubling distinction in kind between human *agents* and nonhuman *actants*, largely because human agents possess the allegedly requisite intentional "self-awareness" to facilitate their exceptional "transformative powers" (2016, 60, 68). While Arias-Maldonado concurs with Bennett that hard-and-fast subject/object distinctions are problematized by "an interfolding network of humanity and nonhumanity," he nevertheless affirms the "primacy of human agency" within the human-nature assemblage (Bennett 2010, 31; Arias-Maldonado 2016, 57–68). Conversely, within that assemblage, Arias-Maldonado understands nonhuman agency to consist in providing the allegedly *stable* material and structural conditions within which transformational human actions may take place (2016, 60). Thus, despite espousing a more relational metaphysics than his Cartesian forebears, Arias-Maldonado retains the same dualistic understanding of human culture as uniquely transformative and nonhuman nature as more or less inert or plastic.

Arias-Maldonado may do so, however, only by failing to properly attend to Merleau-Ponty's, Bennett's, and Plumwood's insights. His human exceptionalism, for instance, is licensed by understanding intentionality in a problematically humanoid manner (i.e., as atomic, self-reflective consciousness). Furthermore, while he eschews Cartesian mechanism, Arias-Maldonado retains a problematic "Newtonian residue." Like Odling-Smee, Arias-Maldonado permits nonhuman quasi agency more or less only to the extent that it minimally conditions (human) evolution in a nonlinear, but entirely *deterministic* (and ideally predictable), manner (2016, 59). More worryingly, it is by crystallizing the terms of this mind/nature dualism in its most extreme manifestation that Arias-Maldonado takes the Anthropocene to provide a better normative lens through which to rethink human-nature relationships.

However, Arias-Maldonado's problem isn't only that he underestimates the mindlike capacities of nonhuman entities. More fundamentally, he appears to misunderstand how actor-network theory must unpack self-aware or intentional human action. For Bennett, as we shall see, intentional actions aren't merely sustained or subtended by their material conditions as Arias-Maldonado suggests (via the slow evolution of microbes which populate human bodies and such). For Bennett, they are, demonstrably, more directly delineated by the quasi-agentic contributions of more-than-human actants (including omega fats and earthworms) in the overall assemblage. Thus, speaking about "human" actions in Arias-Maldonado's radically exclusionary manner relies, for Bennett, upon a neo-Cartesian category mistake; it misunderstands the "human" to occupy a purified and privileged ontological sphere. Arias-Maldonado is, therefore, set dangerously in favor of reducing "*political* agency to human agency" (Bennett 2010, xv) because he takes actions to be definitively human—where humans are discernible as exceptional "drivers" of change—in virtue of their being performed by a human actor. Thus, by adopting the Anthropocene term on the grounds Arias-Maldonado suggests, we risk obscuring nonhuman (quasi) agency and further colluding with dualistic logic in Plumwood's "backgrounding" sense. We also risk something diametrically *opposed* to Bennett's project by using our reflective, linguistic, and conceptual abilities to shift focus *away* from the political agency of *more-than*-human bodies

But Arias-Maldonado's errors aren't simple oversights. They appear to be linked to a wider framework of (meta)theoretical commitments of the ilk that Plumwood problematizes. In the final analysis, like Crutzen, Ellis, and other Anthropocene boosters, Arias-Maldonado uses the misconstruals outlined to ascribe a familiar "transhistorical necessity" to "mastering the natural environment in order to survive and thrive as a species" (Arias-Maldonado 2016, 61). The problem with this recommendation should be clear. It sees more sophisticated forms of anthropogenic mastery as *the* "solution" to the "issues" which constitute the environmental crisis. Not only are some of the objectivistic assumptions behind this project of mastery misled, as itself an integral force in establishing crises, the dualistic drive to mastery cannot underpin environmental theory going forward if we're to attend to our submerged mass.

At the risk of laboring an important point, it is, as Jason W. Moore argues, tacit adherence to this dualistic purview that allows Crutzen and his acolytes to uncritically trace the social roots of the Anthropocene back to a species of practical behaviors—the burning of fossil fuels—rather than the objectivistic and dualistic assumptions which continue to license those behaviors at unsustainable levels. Their "standard narrative" contends that the Anthropocene began when "massive energy subsid[ies]" "locked" in "natural resources" like coal were appropriated by a homogenized, and transsocietal *Anthropos* (Steffen, Crutzen, and McNeill 2007, 616; Moore 2016, 81). As a historical account, Crutzen's thereby normalizes the imperialistic and instrumentalist purview which culminates in the problematically *anti*material mindset of "American materialism." Moreover, by framing the debate as a matter of how humans have exploited and might better exploit natural resources, the standard narrative doesn't "challenge the naturalised inequalities, alienation and violence inscribed in modernity's strategic relations of power and production. . . . It removes inequality, commodification, imperialism, patriarchy and much more from the problem of humanity-in-nature" (Moore 2016, 81–82). By petrifying a lack of critical self-reflexivity to these concerns under the "Anthropocene" banner, its proponents risk legitimating the attitudes and behaviors responsible for our environmental crisis—and, indeed, the Anthropocene's apparently emblematic conditions—in the first place. "There is," as the ecofeminist Chris J. Cuomo puts it, "a unique hubris in characterizing your own culture's destructive patterns, which have been used against other cultures, as definitively human, and then defining those destructive patterns as the harbinger of an already established new-age destined to define the future" (2017, 7).

3.1.5 Back to It

So, where are we? The present chapter attempts to address how a radically reflective ecophenomenological praxis might disrupt the objectivistic purview which precipitates the violence of our environmental crisis. Plumwood and Bennett offer important correctives to our objectifying tendencies by highlighting important subject-object continuities regarding the mindlike qualities that nonhuman entities possess. Their respective identifications of nonhuman intentionality and quasi agency further problematize the reductive terms of scientific naturalism's residual subject/object dualism and expose the implausibility of naturalism's ontologically totalizing approach. Plumwood and Bennett thus help show that natural science's exclusionary assumptions and apparatuses are problematic insofar as they "close off the dialectic" in an asymmetric manner that ensures that one "does not encounter the object as such, in its own fullness" (Haraway 1991, 198; Fox Keller, quoted in Plumwood 1993, 124). If, as Plumwood argues, this superficiality is logically bound to our denigration of the more-than-human world, then these revelations are valuable to a radically reflective rethinking of our submerged mass. Nevertheless, I have also argued that, insofar as it retains a mode of intentional engagement more sensitive to subject-object overlaps than its naturalistic rivals, an ecophenomenological praxis informed by their insights may permit a richer and less exclusionary account of the more-than-human world.

3.1.6 Material Mentality

Where Plumwood and other ecofeminists are often vaguer, however, is on the materiality of subjectivity. This oversight is particularly problematic, given that bottom-up strategies may more directly disrupt the hyperbolically autonomous master-as-subject perspective, rather than shifting taxonomic boundaries to incorporate new incumbents. We've seen how a related failure to problematize humanoid characterizations of nonhuman mindedness limits ecofeminist attempts to rethink subject-object identities with sufficient daring. Plumwood's failure to accommodate the radical heterogeneity of nonhuman intentionality, for instance, was pivotal in my recommendation to resituate her insights ecophenomenologically.

But the problem isn't only that ecofeminists say little about how nonhuman bodies influence the worlds they differentially grasp. In the present groundwork context, investigating the materiality of *human* mentality—particularly intentionality—is essential to unseat naturalism's residual Baconian subject. I argued in chapter 1, recall, that objectivistic ways of revealing the more-than-human world may be traced back, in part, to the attempt to transcend or naturalize one's embodied grasp of the world; to mistake the determinate objects in which perception ends for reality's basic ontological constituents. If the behavioral violence of our crisis situation is as intimately connected to our imposition of conceptual violence as I suspect, then any attempt to reconfigure subject and object (complete with their naturalized inequalities) must investigate bottom-up continuities more thoroughly.

That isn't to say that ecofeminists provide *no* grist for our mill. Warren, for example, is typical among ecofeminists in suggesting that "relationships are not something extrinsic to who we are, not an 'add on' feature of human nature; they play an essential role in what it is to be human. Relationships of humans to the nonhuman environment are, in part, constitutive of what it is to be a human" (1990, 143). However, as Kelly Burns (2008, 102) observes, despite Warren's resistance to mind/nature and subject/object dualisms, she never discusses the salience of one's embodiment or material relationality in any detail. Likewise, Plumwood repeatedly disavows the Cartesian paradigm which sado-dispassionately cleaves mind from body. However, beyond these general commitments, ecofeminists consistently offer little more. Exactly *how* one's materiality saturates one's mentality is usually left regrettably unarticulated.

More troublingly, on the few occasions where Plumwood ventures further details, she is concerned with foregrounding the material dependencies that provide human subjectivity's "enabling condition" (2002, 17). Plumwood thus courts a problematic division of labor similar to that of Arias-Maldonado, which is particularly odd, since her progressive naturalism aims to *show* how "nature as biospheric other is not a background part of our field of action or subjectivity, not a mere precondition for human action, nor a refractory foil to self" (2001, 16). Plumwood *has* given convincing reason not to consider nonhuman entities as mere background instruments in terms of their politico-ethical status. This leaves aside, however, how more-than-human "nature" infuses, dictates, or delineates (human) mindedness to the extent she, rightly, thinks necessary to dismantle the dualistic hierarchism which underwrites scientific naturalism.

So, why the silence? Jensen contends that "ecological feminist aversion toward gender essentialism and biological reductionism seems to have resulted in an aversion to the body itself and has hampered their ability to express an alternative ethic of bodily embeddedness and relation" (2002, 38).[9] Jensen focuses on the salience of embodiment for an environmental *ethic*, however similar concerns about dualistic essentialism seem to be manifested in the absence of a substantive ecofeminist account of how one's material embodiment impacts perception, cognition, and praxis. While we may be sympathetic to these ecofeminists' concerns, their related failure to engage with the *epistemic* salience of embodiment is problematic. In Plumwood's work, as we've seen, this failure allows problematically objectivistic *ontological* assumptions to creep back in the form of a Kuhnian naturalism.

3.1.7 Operative/Act Intentional Intertwinings

Merleau-Ponty, however, fares better. If what I've argued thus far holds, not only are beetles and waste plastic more mindlike than we typically realize, one's reflective consciousness is itself intertwined with a prereflective and intentional worldly bond not entirely unlike those differentially secured by trees or ticks.[10] As Merleau-Ponty argues, operative intentionality is our primary means of access to the world's meanings because one's motor intentionality underwrites the basic grasp upon which epistemological and ontological contentions rely. If the project of recovering nonreductive continuity is essential for better relationships with the more-than-human world, then Plumwood again finds a key ally in Merleau-Ponty. He shows that one's body doesn't just provide the foundation for subjectivity; one's body *is* one's mentality, in a crucial sense.

These contributions may, however, prove beneficial only if Merleau-Pontian ecophenomenology can allay ecofeminist concerns about smuggling a gendered essentialism into any normative account of embodiment. These are important. But since addressing them is the task of chapter 5, I will defer detailed treatment until then. For now, it should suffice to say that there is no obvious in-principle incompatibility between advancing a normative account of one's embodied grasp and maintaining critical self-reflexivity to the sediment which might infuse it. Merleau-Ponty's account of the intentional arc is intended to do justice to this very concern. Moreover, it is attention to sediment that motivates a radically reflective praxis over a fixed, trans-historical, metaphysics. Even in unpacking some of the onto-epistemological significance of one's perceptual embodiment, then, Merleau-Pontians needn't sacrifice a heightened and habitual focus on the kind of critical self-reflexivity rightly encouraged by ecofeminists.

3.1.8 Nonsubjective Subjectivity

Our ecophenomenological investigation may nonetheless benefit from the insights of new materialists who go further, arguing that one *is* one's body, *but so are others*. One may affirm Merleau-Ponty's contention that one's embodied grasp is decisive in revealing the world's meaningful physiognomy. Where we're attuned to our material relationality, however, it isn't "enough to say we are 'embodied.' We are, rather, *an array of bodies,* many different kinds of them in a nested set of microbiomes" (Bennett 2010, 113). For Bennett, the material contributions to one's grasp extend beyond one's functional or physiological orientation. Even the food (or PVC) one consumes (and which becomes, ambiguously, part of oneself) may maintain the requisite quasi agency to shape the world's meanings and our subsequent encounters with it. To illustrate, Bennett cites research indicating relationships between fat consumption and "cognitive dispositions" (2010, 40–41), including proclivity to violence and depression. As we saw in chapter 1, depression may affect one's phenomenological grasp in surprisingly straightforward (e.g., olfactory) ways and delineate the axiological value or affective meanings things present. Ingmar Persson and Julian Savulescu (2012) concur with Bennett's general point. They present research which suggests that elevation of oxytocin levels (a neurotransmitter, the secretion of which is increased by taking certain contraceptive pills and asthma medications) increases our tendency to perceive others as trustworthy. Interestingly, given our exploration of dualism, under increased exposure to oxytocin, subjects' willingness to *sacrifice* individuals of other races (and other "outsiders") also increases.

The quasi agency that Bennett makes explicit also has more banal origins. Earthworms, for example, make soil hospitable for agriculture, thereby coproducing material conditions for the possibility of human agency and certain foods (like those mentioned earlier) that may more immediately contribute to our grasp (Bennett 2010, 98). Moreover, earthworms (and, Ladelle McWhorter [1999] insists, diffuse entities like mineral deposits or aggregations of dirt) may more significantly delineate how we encounter the world by shaping circumstances whereby agricultural resources (or lack thereof) provide motives for action, economic decision making, and prevalent attitudes (e.g., hostilities) toward Others.

But their remarkable (quasi) agency becomes most apparent when it is *disrupted*. Samuel Awuah-Nyameky (2014, 48–49), for instance, observes that anthropogenic climate change has increased the frequency and intensity of bushfires in Ghana's Berekum region. The residual material conditions are largely inhospitable to entities like earthworms which would otherwise facilitate the bush's recovery by improving soil fertility. Subsequent shifts in flora have led many of Berekum's inhabitants to abandon agriculture and take up game hunting instead. More regrettably, from an ecological perspective, local deforestation has also accelerated because newly established hunters actively *set* further bushfires to force their prey out of the bush to be shot. Given that the disruption of nonhuman bodies has radically influenced how Berekum's residents relate to their immediate environments, it is misleading to argue that even human bodies (narrowly conceived) are *the* source of our grasp upon the world which provides our basic means of relating to or behaving toward it.

3.1.9 Problems with Bennett's Naturalism

These important contributions notwithstanding, the ontological flattening Bennett employs is also problematic. One issue emerges when Bennett says that "human actants . . . turn out to *be* confederations of tools, microbes, minerals, sounds and other 'foreign' materialities. Human intentionality can emerge as agentic only by way of such a distribution" (2010, 36; emphasis mine). Understood simply as an attempt to acknowledge our diffuse materiality and the manifold contributors to our behaviors, Bennett's claim has merit. She sharpens radical reflection's critical focus by highlighting some more troubling reasons why perception might have the "anonymous" character Merleau-Ponty claims, for example. But Bennett takes her ontological contentions to have more profound ramifications, concluding that "autonomy and strong responsibility seem to me to be empirically false" (2010, 37). Even setting aside my specific agenda, it's hard to see how such limited autonomy could satisfy *Vibrant Matter*'s explicit "ethical task": to "cultivate the ability to discern nonhuman vitality, to become perceptually open to it," given that we "*are* vital materiality and that we are surrounded by it, even though we do not always see it that way" (2010, 14).

It's tempting to think that I am clutching at straws (as, perhaps, Bennett is) by seeking to retain the degree of deliberative or intentional autonomy to enable one to better grasp the more-than-human world in any important sense. But I don't think so, for reasons that become clear if we scrutinize Bennett's reasoning. Bennett charges her vital materialism with acknowledging the material relationality of ontology in general, and agency in particular. In doing so, she endorses Latour's imperative to disrupt the modernist project of "purification" because that project is irredeemably hostile to relationality. However, like Latour, Bennett cannot fully satisfy its demands because she herself attempts a purification of the *epistemic* relations between situated, embodied, beings and the worlds they theorize. The disconnect between her attempts to problematize hyperbolic autonomy or transcendental subjectivity (by appealing to the surreptitious contributions of omega fats and vitamin levels), and her totalizing description of ontological reality-in-itself (as definitively elucidated by a curiously desituated actor-network theory) is problematic. And, in the final analysis, it is only by flattening ontology according to the residual assumptions and prejudices licensed by objective thought that Bennett's austere conception of autonomy is demanded.

It is telling, for instance, that Bennett's rejection of autonomy and responsibility rests upon their *empirical* falsehood. We've explored empiricism's problems at length. As a manifestation of objective thought, empiricism is problematically reductive and inattentive to the perceptual fields from which its insights derive. Of course, Bennett is no straightforward empirical realist in the Baconian or Kantian mold. Nonetheless—as her repeated attempts to describe *the* fundamental contours of mind-independent reality-in-itself demonstrate—she retains a related commitment to metaphysical naturalism which is, at base, similarly constrained by objective thought's stipulations and predilections. If Bennett retains the naturalistic hierarchy of explanation under which empirical descriptions of "objective" reality retain deliberative priority over more straightforwardly perceiver-dependent phenomena (such as greater deliberative or intentional autonomy) revealed phenomenologically, this would help clarify why Bennett permits only the austere (and homogenized) autonomy she does. However, Bennett's naturalistic commitments are especially problematic if, as Merleau-Ponty argues, experience is the *only* "ultimate court of appeal in our knowledge of things." Given Bennett's opposition to modernist purification, this is an insight to which she should be amenable.

Independent of the push toward naturalistic realism (which perhaps imbues Bennett's "materialism"), and heeding Plumwood's decree to acknowledge continuity *and* difference, it seems to me that one must admit the overall inescapability of relational selfhood that Bennett exposes. However, one needn't *reduce* selfhood—nor one's agency or autonomy—*to* those relations in the manner Bennett implies. Bennett rightly encourages attentiveness to one's affective and behavioral entanglements with bodies beyond one's epidermis. By showing that the inherently unpredictable more-than-human world cannot be satisfactorily subsumed under any totalizing schema, she also helps keep open the dialectic of human-world engagement. Yet, given our renewed attention to alterity and heterogeneity, Bennett's cannot be considered an exhaustive account of ontological reality entirely independent of the horizons and apparatuses from which it emerges. By resituating Bennett's insights ecophenomenologically, however, we're given more reason for cautious optimism about the possibility of positive change than Bennett herself consistently maintains.

Nonetheless, emphasizing the somewhat ambiguous diffusion of mentality is important. A greater *attentiveness* "to the indispensable foreignness that we are" not only might help disrupt dualistic hierarchism and objectivistic metaphysics, but also might disrupt the entrenched tendency "to produce and consume in the same violently reckless ways" (Bennett 2010, 113). Why? Because the pervasive fantasies of human mastery diagnosed by ecofeminists are shown more clearly to be unachievable, dangerous fantasies: "In the most obvious sense, if one cannot presume to master one's own body, which has 'its' own forces, many of which can never be comprehended, even with the help of medical knowledge and technologies, one cannot presume to master the rest of the world, which is forever intra-acting in inconceivably complex ways" (Alaimo 2007, 250). One thing Alaimo means to suggest here is that critical self-reflexivity about one's material embodiment may direct one away from a project of conceptual and behavioral colonialism and toward a stance of *humility*. Given that he understands humility to be exemplified by commitment to the sort of onto-epistemological openness facilitated by a praxis of radical reflection, Alaimo's sentiment could have come straight from the pages of Merleau-Ponty.

3.2 SEEING BETTER

Let's recap. I've argued that ecophenomenologist, ecofeminist, and new materialist insights problematize the dominant subject/object terms, and related metatheoretical assumptions about "possession," that contribute to the "submerged mass" of our environmental crisis. Where one can suspend fundamental commitment to the objectivistic metaphysical stipulations of our default naturalism, I've suggested, one may be open to a richer range of phenomena (e.g., nonhuman intentionality, inorganic quasi agency, or axiological meanings) than can be consistently sustained within the ordinary remit of natural science. This is important if, as I've argued, naturalism's objectivistic commitments precipitate our crisis situation by legitimating the violent attitudes and behaviors responsible for its apparently constitutive "environmental issues." By engaging with the more-than-human world beyond these confines, however, the subject-object continuities identified above undermine any exhaustive project of ontological or epistemological objectification. Simply put: (nonhuman) intentionality and (quasi) agency prevent the world from being reduced to a determinate or determinable collection of *objects*, properties, or relations-in-themselves as naturalism ordinarily demands.

I've also recommended that the aforementioned insights be reconfigured ecophenomenologically. This is because ecophenomenology is uniquely well situated to evade the residual objective thought which pervades even the "suitably adjusted" naturalisms pursued by Plumwood and Bennett. Contrary to appearances, these naturalisms may retain traces of the subject/object dualism through their failure to address the radical contribution of (other) bodies (otherwise significantly confined to the "object" realm) to one's underpinning epistemic grasp (implicitly: the privileged realm of transcendent or transcendental "subjects"). I've argued that critical self-reflexivity about such concerns is essential to maintain openness to heterogeneous more-than-human entities and their radically differential worlds.

The above arguments effectively amount to the requirement to deny natural science a monopoly on ontological and epistemological matters where the more-than-human world—and, particularly, the environmental crisis—is concerned. To reiterate: this is because natural science's totalizing project requires substantial ontological purification of the "objects" of its concern from the interrogative "subject's" means of epistemic access to them, a task rendered implausible and politically suspect by the insights previously surveyed. If I'm right about this, however, then we must replace the naturalist's God's Eye view with a norm which takes more seriously that the world cannot be unproblematically objectified from an acosmic standpoint. To appreciate how Merleau-Pontian ecophenomenology may provide such a norm in the form of a "view-from-everywhere," however, we should first shift gear and retrace a key insight from chapter 1.

3.2.1 Seeing Things

I've suggested that Merleau-Pontian ecophenomenology's unique ability to resist objective thought comes through its observation that perceptual content is positively indeterminate. For Merleau-Ponty, reality is expressed through a Gestalt structure which draws upon the world's normative impetus and the contribution of the situated body-subject. In this Gestalt context, meanings are determined by their diacritical relationships within an overall perceptual field.[11] Even a carpet's redness or the odor of mammalian blood, Merleau-Ponty argues, cannot be entirely separated from one's differential bodily expectations, which reflect one's multifaceted intentional relations with the world. For Merleau-Ponty, perceptual background is, therefore, normatively salient to the meaningful physiognomy of the intentional objects in which perception culminates, because the background "directs my gaze and causes me to see the object" (*PP*, 310) as I do. Background *is* perceived as a field-quality, if not as a determinate "thing" in its own right.

This fundamental insight is ordinarily missed when we grant naturalism its privileged onto-epistemological status. Naturalism backgrounds its own diacritics—its (meta)theoretical assumptions, intentional commitments, projects, and so on—and justifies this move through an insular dialectic which limits ontological reality to the modality of the in-itself. Now, although his project lacks an environmental focus, the seeds for a more positive onto-epistemological norm can be located within Sean Kelly's (2004) exploration of how the aforementioned figure-background relations permit "object" perception at all. I shall, therefore, first explore Kelly's account before addressing its ecophenomenological repercussions.

Kelly's goal is to unpack Merleau-Ponty's account of how humans perceive substantive "objects" (here: any ostensibly discrete, three-dimensional entity) and, more importantly, how we come to perceive them *better,* despite our partial perspective. His main example is of an old Western film set. Upon entering the set, one might see certain buildings: the saloon, jail, and the like. After exploring the set, however, despite the same "objective" visual panorama presenting itself (i.e., certain frontages with closed doors), the set looks—indeed, *is*—different. One now perceives facades rather than buildings because one's failed exploration of the saloon's and jail's interiors frustrates one's expectations, inaugurating a Gestalt-shift.

Kelly's point isn't simply that one's perception here *changes.* Given that the revised perception is more accurate, evident *progress* has been made. Which isn't to imply that one has merely acquired the requisite sense data to represent some discrete thing in a manner which better corresponds with its own roster of nonrelational qualities. Merleau-Ponty's arguments against objective thought put paid to any such conception. For Merleau-Ponty, Kelly thinks, we register progression because, in perceptual acts, one is always already implicitly involved with an optimum which, in the case discussed, one has moved toward. Moreover, Kelly contends, the derivative quality of one's situated perspective from this optimum is phenomenologically available.

Kelly's sounds like a strange claim, but some compelling cases support it. As a human (rather than, say, arachnid) body-subject, there *are* increasingly optimal distances to stand from a painting to grasp its detail or appreciate its chiaroscuro, which we seek by shifting position in its vicinity. Likewise (unless we share canine chromatic limitations, for example), there are increasingly optimal lighting conditions under which to discern what we almost irresistibly take to be a fabric's "true" color; a phenomenon psychologists track via the well-established "color constancy hypothesis," and which explains our frustration with the poor lighting in department store changing rooms. In each case, one's embodied gaze appears to be drawn by the normative background to a more accurate or sharper presentation of the "object" in question. As Kelly puts it: "I see how the lighting *should* change in order for me to see the color better" (2004, 83).

Kelly emphasizes the intimate, tripartite relationship between figure, background, and one's body involved in any progression toward perceptual optimality. Take our example of color perception. One is only ever "objectively" party to situated, perspectival, instantiations of lighting and colors, yet one unavoidably perceives the constant, "*real*" color, differently situated.

This forms the background of further experience with it (e.g., as "distorted by poor lighting" or "overlaid with shadows"). But, for Merleau-Ponty, such normativity is intimately linked to one's embodied *grasp*, given that perception's differential intentional character reflects one's comportment into the world and the guiding expectations therein. For Merleau-Ponty, Kelly suggests, these optimal background norms must, therefore, be accessed prereflectively as something like the *maximum* grasp one *might* have on the feature in question. For his part, Merleau-Ponty contends that "The real colour persists beneath appearances as the background persists beneath the figure, not as a seen or thought-of quality, but through a non-sensory presence" (*PP*, 305). Which is to say that the "real color" is implicitly determined with reference to "the color that is optimally illuminated by the lighting norm, and this lighting norm is never determinately experienced" (Kelly 2004, 88).

It may be useful at this point to curtail some potential misunderstandings regarding the appeal to "real color" with implications for my ecophenomenological reworking. For reasons outlined above, Merleau-Ponty cannot determine an object's "real" meanings by appealing to their discrete existence wholly antecedent to perception, nor to some ready-made thing-in-itself in which they latently inhere as determinate properties. But, according to Kelly, neither does Merleau-Ponty take meanings like something's real color to be simply conventionally or differentially expressed through contingent relationships between perceiver, thing perceived, and contextual lighting conditions, which would fail to do justice to the peculiar constancy phenomenon with which the relevant section in the *Phenomenology* is concerned.

In Kelly's reading, lighting context *is* inseparable from color because the former contributes to the background against which the latter may emerge as a graspable phenomenon (i.e., *as* a color) at all. However, the constancy hypothesis implies that lighting remains normative, albeit only marginally. Thus, although color perceived and lighting context are partly constitutive of a "unified structure" which also makes reference to our situated bodily contributions (hence the caveat about canine vision), "the lighting context figures in experience by leading my gaze to the optimum place where the lighting best presents the color" (Kelly 2004, 85). And it is because of their interdependence in situated perception that both lighting context and the color correlative to it are experienced as derivative from the optimum—the color that would be optimally illuminated by the lighting norm (i.e., the real color)—which is nonetheless implicitly present (but never *determinately* experienced) in concrete perceptual instances (Kelly 2004, 88–89). Phenomenal optima like real colors, it seems, may be both intrinsically relational and context-sensitive, without sacrificing their peculiar normativity.

Despite this concession to interdependence, Kelly's interpretation might remain uncomfortable for those who want to claim that, despite his interest in the constancy hypothesis, Merleau-Ponty isn't committed to the existence of "real" colors at all. The contention that color merely expresses itself differently in relation to different lighting conditions might be supported with reference to the essay "Indirect Language and the Voices of Silence," in which Merleau-Ponty differentiates "creative" and "derivative" expression. The former "speaks peremptorily when it gives up trying to express the thing itself," establishing "a field of expression that, once appropriated . . . begins to generate meaning out of which its significations can be appropriated by others" per the latter (*ILVS*, 44, 46; Silverman 1980, 132). Kelly's interpretation might be threatened if we should understand talk of supposed constants to be derivative of the idiosyncratic creative expressions which establish any such meanings. The (potentially heterogeneous) institution of meanings like color might then rule out the temporal or ontological priority that Kelly's talk of a definite, "real," optimum implies.

However, this concern appears to me misplaced for reasons explored in chapter 1. There, we saw that radical reflection focuses partly upon what David Morris (2018, 129) calls "proto-sense": the debt of meaning that we owe the world with which we coconstruct our basic intentional structures, and which remains integral to the expression of phenomena. Which is to say that the world itself licenses certain meanings which show up phenomenologically, even if those meanings—*qua meanings*—cannot preexist the creative expressive contexts in which they emerge. The sky's irresistible blueness, for instance, cannot exist *determinately* over and above its differential expression. But this concession needn't undermine the world's contribution to those idiosyncratic creative manifestations via otherwise indeterminate, but nonetheless normative, *grounds* of them.[12] Even creative expression, Merleau-Ponty writes, "is a response to what the world . . . demanded" (*ILVS*, 59). And so it must be, otherwise there would be no obvious reason for our unshakeable conviction that the sky really is blue (for those with visual systems attuned to seeing it).[13] "Real color," it seems, may show up only within situated, meaningful expressions, but does so marginally within their normative background and hence remains consistent with (and instructive to) Merleau-Ponty's wider account of expression.

Having addressed concerns about the interdependence of phenomena and the ontological priority of meaning, we may safely return to Kelly's main example of seeing whole *things* (rather than properties thereof). As Kelly (2004, 91) clarifies, however, there is a relevant disanalogy here because the optimal position to observe three-dimensional "objects" cannot be even ideally instantiated. To see the film set optimally, for instance, is to see it from *everywhere*, because each perspective reveals some otherwise unavailable feature (*PP*, 69). A literal view-from-everywhere is, however, impossible for any situated perceiver to possess, and all perception is irreducibly situated: "I say that I perceive correctly when my body has a precise hold on the spectacle, but that doesn't mean that my hold is ever all-embracing; it would be so only if I had succeeded in reducing to a state of articulate perception all the inner and outer horizons of the object, which is in principle impossible" (*PP*, 297). Kelly contends, however, that "although it is not itself achievable by me, the view-from-everywhere is nevertheless an ideal from which I sense myself to be deviating in acts of perception. It is the *norm* . . . with respect to which all actual points of view are understood" (2004, 92). Insofar as one may perceive a substantive saloon-object, for instance, one's perception is already partly constituted by reference to its hidden sides in a similar way that a fabric's redness refers to how it would appear under optimal lighting conditions.

Carl B. Sachs therefore argues that, for Merleau-Ponty, three-dimensional objects are perceived "in full" because their "non-facing sides are perceptually present to us in the mode of [counterfactual] *possibilities*" (2017, 92). One's perception of the saloon includes how it would look if one somehow "took up" the jail's position, for instance. Kelly (2004, 80–81, 100) rightly argues, however, that any such "possibilities" cannot merely be how the hidden sides would look *if I* went and stood there (as Husserl claims), because those other perspectives are manifest, albeit indeterminate, aspects of my current perception. They are partly constitutive of what it is to *see* (rather than, as Husserl suggests, hypothesize) three-dimensional things. Kelly's claim is exemplified in the following passage from Merleau-Ponty: "To see is to enter a universe of beings which *display themselves*. . . . To look at an object is to inhabit it, and from behind this habitation to grasp all things in terms of the aspect which they present to it. But insofar as I see those things too, they remain abodes open to my gaze and, being potentially lodged in them, I already perceive from various angles the central object of my present vision" (*PP*, 68). The present perspectives of other entities (in which one may be lodged, and which are licensed by the "object"), partly delineate the normative background of the intentional objects that perception reveals, because "every object is the mirror of all others" (*PP*, 68). For Merleau-Ponty, Kelly argues, the *view-from-everywhere* is the implicit optimum from which all factual perception is derivative. By better satisfying this norm's demands (minimally: by appreciating how the saloon's hidden sides look) we perceive things *better*.[14]

3.2.2 The View-from-Everywhere

How might the foregoing help to establish the view-from-everywhere's credentials as a nonobjectivistic onto-epistemological norm? To answer this question, it's important to appreciate that Kelly limits himself to exploring how *objects* are revealed as three-dimensional wholes to *human* perceivers, through our "full bodily readiness" (2004, 101) to navigate them: more or less how nonhuman things show up substantively in our everyday utilitarian praxis. Kelly says little, therefore, about how more-than-human entities might resist this sort of objectification when differently engaged. Nevertheless, Kelly's basic account of the normative character of phenomenal experience rings true. Moreover, I think we can extend his analysis to accommodate other kinds of ontologically real meanings that the preceding analyses legitimate.

To explain: both Kelly and Merleau-Ponty appear to take the view-from-everywhere to refer to the *norm* determined by the sum of all interior and exterior horizons under which something might be truly expressed.[15] I've argued that no objectivistic mode of revealing could wholly satisfy its demands since, even insofar as such a purview's revelations may be true, its apparatuses and assumptions exclude other meanings plausibly licensed by the more-than-human world itself. Uexküll's ticks, for instance, don't obviously experience raisins *as* intentional objects. It is misleading (and somewhat colonial), therefore, to take the phenomena *we* designate as "raisins" to be exhausted by the multitude of raisin-object perspectives, even if that's how they ordinarily appear for us. Doing so would reinstate the prejudices licensed by objective thought because the phenomena in question would be effectively reducible to the perspectives *like mine* in which they may differentially feature as objects.

Given that the above error is paradigmatically Kuhnian, it's unsurprising that things are similar in the scientific context. Atmospheric chemists like Crutzen offer observations about greenhouse gas emissions which are, for the most part, evidently true. Moreover, their observations are *essential* to understand some plausible causal relationships between fossil fuel consumption, desertification, and ecosystem collapse. Crutzen's discipline may also help mitigate anthropogenic climate change by modeling alternative resource allocation strategies, cleaner fuel sources, and the like. However, two problems prevent us from taking climate change or its constituent aspects to *be reducible to* narrow causal relationships between the determinate extensional properties of atmospheric gases, ultraviolet radiation, multicellular organisms, and so on. First, by taking science's positivistic remit to exhaust the boundary conditions of ontological and epistemological enquiry, natural scientists *cannot* register the heterogeneous meanings or horizons surveyed in section 3.1 which violate its more fundamentally objectivistic demands. Second, by mistaking its reflective abstractions to be ontologically primitive, science limits or distorts the (allegedly determinate and wholly determinable) phenomena it can. This issue cannot be resolved by taking multiple scientific perspectives to exhaust the phenomena in question, as our investigation of Warren's hierarchy theory in chapter 2 illustrates. To reprise an earlier insight, the most we can say is that, insofar as Crutzen's observations ring true, their truth cannot be entirely disentangled from their background projects, assumptions, and apparatuses.

It is, therefore, both false and deeply problematic to mistake the various regional truths natural science yields—either individually or collectively—for an exhaustive account of (a given aspect of) the more-than-human world. I say "deeply problematic" because, by taking the scientist's phenomena to *exhaust* the world's furniture, one risks naturalizing science's more troubling utilitarian or dualistic assumptions, which are incompatible with disrupting the environmental crisis' underpinning violence. The point isn't that natural science's contentions aren't true, but that other intentional horizons *really* may yield different meaningful phenomena. Only via a problematic commitment to naturalism's objectivistic prejudices may one outright deny this richer range of phenomena their place in any view-from-everywhere.

Now, if one really were to somehow perceive all the interior and exterior horizons of a rainforest, say, one would perceive all the meaningful expressions that entity truly motivates, and any meanings which it may *itself* express.[16] Exhaustively articulating these would be impossible, but here's a selection: from certain economic perspectives, the rainforest might be a fuel or capital resource, perhaps with axiological meanings which are "objectively" incompossible with those available from alternate viewpoints. As Charles S. Brown notes, "clear-cutting large tracks of old-growth forest may appear good from the perspective of business and profit whilst appearing evil from the perspective of wildlife habitat preservation" (2003, 11). These axiological meanings, of course, cannot show up under Crutzen's schema because they cannot be registered by the extensional realist apparatuses which regulate objectivity within that field. Our list, however, must also include more radically different meanings. "Taken in itself—and as an object it demands to be taken thus" (*PP*, 70), a view-from-everywhere should, by right, include the differential expressions the rainforest motivates, including those configured by otherwise engaged humans, orangutans, individual trees, compound forest ecosystems, earthworms, and perhaps even minimal cellular entities.

This list would be long and, in fact, articulating *all* the rainforest's meanings is impossible. I've argued that, given our intractable situatedness, *some* meanings must be unavailable to us on their own terms, such as the *entire* intentional reality of the dog, tick, orangutan, or tree. One may have a degree of epistemic access to the realities of more-than-human Others with which one is connatural. Nevertheless, taking alterity seriously precludes full transcendental access to them. Just as I cannot wholly know how "raisins" (don't) show up for ticks, I cannot wholly know how the "rainforest" shows up for its primate inhabitants or other situated body-subjects. Nor can I legitimately or consistently exorcise those meanings from the world. This is why Merleau-Ponty calls individual expression only "a certain accent of the world's style" (*PP*, 406). I address the issue of alterity more fully over the next three chapters. It will suffice to say for now that, under any anticolonial view-from-everywhere, there must be a *revealing* excess which a situated body-subject cannot grasp *except as an excess* without doing undue violence to alterity.

But such an excess shouldn't be admitted to our view-from-everywhere only to acknowledge other intentional realities. Taking Bennett's insights seriously, a further excess is required to acknowledge the (quasi) agency retained by the trees, rainforest ecosystem, its nonhuman inhabitants, and so on. We can better appreciate Bennett's point in the context of another "issue" associated with the environmental crisis: coltan mining.

Coltan is a metallic ore used in manufacturing surgical equipment and mobile telephones. The commonplace practice of pan-mining coltan, however, produces toxic by-products which, in developing countries like the Democratic Republic of Congo, are typically left to pollute water sources and destabilize local ecosystems. Coltan mining is also directly implicated in deforestation, the destruction of stream beds, the collapse of sustainable local agriculture, and the hunting to near extinction of mountain gorillas

(Bleischwitz, Dittrich, and Pierdicca 2015, 159; Plumptre et al. 2002, 420–21). The final issue arises because coltan mines are often far from existent food sources and so (to borrow markedly technoscientific terms) huge quantities of "bushmeat" "fuel" its excavation. Even setting aside many of its other aspects which cannot be captured "objectively," scientific descriptions typically err by taking concrete instances of coltan mining to consist of *ideally predictable* relationships between physico-chemical properties or systems. Understood as such, the negative effects of coltan mining would be entirely foreseeable (and, presumably, largely mitigable) with the assistance of a sufficiently sophisticated set of sciences.

There's a kernel of truth in this thought. If science had been in the same position in the 1780s as it is now, we'd doubtless have a better idea about the likelihood that anthropogenic climate change and ecosystem collapse might result from shifting from agrarian to industrial societies. It's also possible that we might have legislated differently for subsequent industries like coltan mining. Nevertheless, if Bennett is on the right lines, the dominant scientific picture is dangerously simplistic because it cannot acknowledge the "thing power" possessed by its contributory bodies and the collectives they form. In fact, the reason I mention coltan mining is that a closely related situation *did* lead, in a not wholly predictable manner, to the emergence and globalization of HIV-1.

In the early twentieth century, Congolese nationals under Belgian colonial rule were conscripted to work in rubber plants or to build railways for the transportation of goods. A combination of enforced monoculture (land was taken from crop farmers to satisfy rubber demands) and huge nutritional requirements (thousands of migrant workers built the railways) jointly ensured food scarcity. This industrial hub then incubated a mosaic of interconnected factors which seem to have jointly led to the transmission of HIV-1: malnourished workers who hunted chimpanzees; unsafe food preparation methods; a prevalence of chimpanzees infected with Simian Immunodeficiency Virus (SIV); SIV's more discrete quasi-agentic capacity to evolve into HIV-1 and permit human transmission; an increase in commercial sex workers; and a newly constructed, large-scale transportation system, to name a few.[17]

The similarities between situations are striking. So, if, as Bennett would surely agree, the globalization of HIV-1 is more or less inexplicable without reference to quasi agency, then the same might be said about the present coltan situation. Worryingly, we can reach this conclusion without mentioning the profound repercussions that scientists say probably *will* result from ecosystem collapse or water pollution in those regions. Taking Bennett's contentions seriously, then, I submit that the agency retained by waste deposits, slime molds, power plants, crystals, and HIV-1 *must* show up as a *real* excess in any account of these things where anthropogenic mastery is decentered.

What's the moral of my story? I've argued that, since the more-than-human world cannot be unproblematically confined to a purified order of "objects," the naturalist's God's Eye view should be replaced with a view-from-everywhere. However, a view-from-everywhere sufficiently self-critical to disrupt our imposition of conceptual violence cannot limit itself to the truths revealed by multiple scientific or objectivistic schemas. Insofar as these schemas may reveal truth, their reductive stipulations obscure some of the meaningful phenomena available for expression, some of which cannot be thematized without distortion, except as an excess. Nonetheless, these excesses must be retained if, as Kelly suggests, seeing *better* involves more fully appreciating the various meanings that an entity itself permits.

Adherence to this revised view-from-everywhere means that the more-than-human world cannot be ontologically or epistemologically *possessed* in any totalizing manner. It cannot, after all, be reduced to natural resources, ecosystem services, or any of the "objects" which show up within a given intentional horizon. Nevertheless, I've argued that the failure of a project of mastery shouldn't give cause for regret. It should serve to inculcate a radical sort of critical self-reflexivity about our imposition of conceptual and behavioral violence. According to the line of argument I've been pursuing, it is principally by reducing one's imposition of epistemological and ontological violence that one may *better* perceive and relate to the more-than-human world. By contributing to a shift from mastery to humility, ecophenomenology might also thereby help disrupt our "submerged mass."

Of course, one may have worries about my appeal to the view-from-everywhere in pursuit of this goal. The first pertains to concerns about its apparent proximity to the Baconian purview it intends to correct. Proponents of this concern may worry that determining ontological reality with reference to the sum of *all* perspectives is tantamount to the pursuit of the transperspectival God's Eye view, or the modern era scientist's Laplacean aspiration to see all the causes so as to determine the future more accurately. The main risk here is that "vision, in its critical function of revealing hidden conditions and making visible the secrets of visibility, become[s] totalizing" (Al-Saji 2009, 384).

These worries would, however, be misplaced because the view-from-everywhere is not totalizing in the problematic sense. *My* view-from-everywhere aims to address meanings differentially expressed by entities beyond a narrowly humanoid perspective. However, unlike the Baconian or Laplacean purview, it's also geared to remain cognizant of its intractable bodily and social-situatedness and thus demands critical self-reflexivity about the meanings that we, situated body-subjects, locate within it. This concern led me to emphasize the manifold (nonobjectifying and thereby, in Alia Al-Saji's sense, *in*visible) excesses which must punctuate it in recognition of those meanings which are unavailable to me as a situated body-subject (vs. Bacon), and the quasi agency, for instance, which undermines the Laplacean causal story. Likewise (as Kelly's film set shows), it is also why the view-from-everywhere must remain, as Merleau-Ponty says, "incomplete" and "presumptive" (*PP*, 70): subject to correction and reconfiguration as experience and radical reflection demand. In an ecophenomenological context, such provisionality is pivotal to the humility thereby inculcated. If my appeal to the view-from-everywhere *is* totalizing, then that term applies only to its widened scope of concern rather than to any pernicious attempt to possess or control the more-than-human phenomena therein.

A second worry pertains to the compatibility of my conception with Merleau-Ponty's, given that, when discussing "the house seen from everywhere," he remarks: "The completed object is translucent, being shot through from all sides by an infinite number of present scrutinies which intersect in its depths leaving nothing hidden" (*PP*, 69).[18] The concern here is that an *infinite* number of perspectives cannot result in the *sum* of all meaningful perspectives and that seeing better risks becoming impossible insofar as it relies upon doing so.

One way to overcome this issue is to problematize Merleau-Ponty's claims about an infinite number of perspectives. One might emphasize what I said earlier about the world's normative role as grounds in delineating the meanings that show up phenomenologically, in virtue of which worldly entities might motivate only a finite, but vast, number of differential expressions. Alternatively, one might emphasize that the appeal to an infinite number of perspectives is as problematic for Merleau-Ponty as it is for me, if one's present perception of the house is meant to include awareness of them.

For my part, I remain somewhat neutral on the issue. I'm tempted, on the basis of the first strategy, to deny an infinite number of perspectives.

However, one needn't make this move. Seeing *better* (and not *perfectly*, which, for reasons already explored, cannot be one's goal) demands increased openness to multiple perspectives, rather than correspondence with a definite viewpoint. I don't think seeing *better* would, therefore, become impossible even if an infinite number of meanings were available for expression. Whichever position one adopts, I've repeatedly emphasized that, although normative, a complete view-from-everywhere is *unobtainable*, and this admission is important in developing the theoretical and behavioral humility characteristic of the ecophenomenological praxis I advance. Unpacked as such, an infinite number of possible perspectives upon the house would prove *no more* problematic for Merleau-Ponty either, since the view-from-everywhere manifested in a particular expression of the house remains necessarily (and thankfully) incomplete and subject to revision in view of further (not *all*) perspectival insights.

Finally, regardless of whether one takes seeing better's improvement curve to be asymptotic, one might be skeptical about there being *one* true meaning over and above particular expressive acts (i.e., *the* view-from-everywhere). Proponents of this concern might again worry about such a meaning's consistency with Merleau-Ponty's doctrine of (e.g., "creative") expression and the attention to heterogeneity emphasized elsewhere.

However, as I noted when exploring Kelly's reading of Merleau-Pontian color perception, the point about phenomenal optima isn't that there is one true meaning, somehow individually possessed by each thing (its definite, nonrelational, color, say), which each phenomenon subsequently expresses. As Morris says, despite relying upon "a logos in things that grows out of them . . . primary expression is a creative endeavor that does *not* begin from or express something already fully given or determinate, yet expression precisely engenders ways of making sense that *do* grasp things as given and determinate" (2018, 125, 166). The point, for me at least, is that the different meanings by which an entity may be truly expressed are motivated by the grounds that entity provides and contribute to the various ways it might be identified and individuated. Such grounds permit awareness of the carpet's optimal redness for those physiologically attuned to them, for whom this norm shows up indeterminately in perception. But *our* determinate, veridical, perceptions remain qualitatively heterogenous with others', more so with dogs and so on, for whom the grounds which delineate "real color" are presumably manifested rather differently. Likewise, a raisin or rainforest might be configured in manifold meaningful ways, yet, partly because of their somewhat common grounds, these meanings might be variously unified as "accents of the world's style," to which the view-from-everywhere aims to do justice. The view-from-everywhere, in short, aims not at univocity of meaning or hostility to difference, but at increased openness to both. This is the main advantage of my view-from-everywhere over Kelly's. It would be more conducive to what Al-Saji calls "the project of critical ethics": aiming not "to see others in exhaustive (and hence inevitably reductive) ways, but to see according to other ways of seeing, elided in the logic of objectifying vision; it would see according to, or with, others" (2009, 385).

3.3 BETTER HABITS

In the broadest terms, I've argued that Merleau-Pontian ecophenomenology is particularly well situated to address our imposition of violence because of its demand for radical critical self-reflexivity about the various implications of being a situated body-subject. But, in the foregoing discussion, I left an important thread unattended. Taking embodiment seriously *at all levels* also requires acknowledging that we learn to perceive and relate to the phenomena we encounter as a matter of embodied *habit*. In chapters 1 and 2, I argued that identifying sedimented perceptual habits is important for understanding why, even as environmental philosophers, we enframe the world under misleading objectivistic or technoscientific terms. The root problem, I suggested, isn't the pervasive influence of scientific ontologies and epistemologies per se, but rather that, in its role as *the* privileged discourse, natural science legitimates a problematic kind of objectification partly underpinned by dualistic sediment and a related adherence to the metaphysical prejudices licensed by objective thought. Both are, in significant part, *learned* ways of perceiving which have come to predominate, to which we nevertheless remain significantly tied—at an operative level—as a matter of habit.

Emphasizing that perception is habitual and imbued with the sediment of one's multifaceted situation is important. In chapter 2, I recommended reconsidering many ecofeminist "solutions" to dualism because dualism cannot plausibly be overcome by mere assent to a better conceptual scheme. This admission might prove similarly problematic, however, for my recommendation to take up a less objectivistic or dualistic view-from-everywhere. One cannot just begin to perceive and relate to the world better because, as Kohák argues, objectifying and commodifying the more-than-human world is a staple of our habitual praxis. Therein lies the limitation of radical reflection where it is conceived only as a conceptual tool. While, within one's "urban" situation, one might become aware of one's imposition of onto-epistemic violence in rare moments of reflective clarity, given that this is when one is often *least engaged* with the world, these are unlikely to impact on our attitudes or behaviors in any profound or lasting manner. And these are the sorts of changes we need to effect to address our crisis situation at root.

So, the residual thread is as follows: regardless of any formal commitment to a revised view-from-everywhere, how can one prevent oneself from habitually enframing the world in objectivistic and utilitarian ways? The answer, I suggested in chapter 1, must lie in developing the means to sustain a *praxis* of radical reflection: to instill the requisite habitus of critique to identify and resist the violence done in normalizing or naturalizing objectivistic ways of revealing. In this chapter's final section, I entertain some suggestions about how to do so.

3.3.1 Attention

On the face of things, the answer is simple. One must cultivate the virtue of *attention* to the things themselves. Plumwood's and Bennett's promotion of openness or care as improved modes of epistemic engagement was motivated by the attempt to do so. Merleau-Ponty makes similar remarks when he contends that phenomenology is epitomized by its "painstaking . . . kind of attentiveness" and the "demand for awareness" that motivates our openness to "*being filled with wonder*" by the phenomena themselves (*PP*, xiv, xxi). For Merleau-Ponty, ongoing attentiveness facilitates radical reflection. The habitual inattentiveness cultivated under dualism, for instance, frustrates it. These remarks aside, our question then becomes: How can one cultivate habitual attentiveness in the requisite manner?

Bennett's account of "American materialism" suggests that our objectifying tendencies may be licensed partly by an unreflective familiarity with things. Her contention is entirely consistent with the analyses in preceding chapters. Where one is sufficiently familiar with entities intersubjectively characterized as "tools" or "use-objects" (whether technoscientifically, or in an everyday consumeristic context), it seems that their utilitarian meanings may be sedimented into one's intentional arc. But this gives us a problem. Since our bodies are our medium for having a world at all, these sedimented contributions to the diacritics of perception don't reveal themselves among the objects we experience, nor are they straightforwardly transparent to reflection. Al-Saji puts this concern as follows: "We see *through* our habits, through our eye movements and bodily kinesthesis; we do not see them. This self-erasure is what allows habits, once acquired, to appear effortless and 'natural'" (2009, 378). As with the blind person accustomed to her cane, familiarity with our "tools" *distracts* us from attending to our appropriation of them, because our unreflective familiarity "*relieves us of the necessity of doing so*" (*PP*, 364). In *Humanism and Terror,* Merleau-Ponty expressly links this process of instrumental appropriation with colonialism: "Consciousness can do nothing without its body and can only act upon others by acting upon their bodies. It can only reduce them to slavery by making nature an appendix of its body, by appropriating nature to itself and establishing in nature its instruments of power" (*HT*, 102).

Revelations like the above lead Erazim Kohák (1984, 23) and Simon P. James (2009, 152) to recommend distancing oneself from the artifactual world. While the former literally prescribes fleeing to (and remaining in) the woods, the latter recommends periodic breaks from "merely functional" entities because what James calls "encounters with nature . . . can help one develop the attention which is the mark of phenomenological inquiry done well" (2009, 105–8).[19] This is because the least markedly artifactual things—unlike consumer goods like cars or mobile telephones—aren't *so* finely tuned or ergonomically crafted as to quietly satisfy our use for them. By surrounding ourselves with things that resist our appropriation of them, the argument goes, we may disrupt our objectifying habits sufficiently to afford the requisite epistemic space in which things begin to hold our attention. We might spend time surrounded by cedar trees, for instance, which might satisfy some fuel or engineering needs, but not without significant manual labor. Nor, in this face-to-face setting, will they typically disclose themselves *wholly* as "natural resources." This sort of functional break might serve to cultivate ecophenomenological attentiveness.

We might nevertheless have some reservations about James's and Kohák's specific proposals. The first emerges in the context of Bennett's rubbish: namely, that one must afford the inorganic or "functional" world the same attention as imposing Sequoias, glaciers, or mountain ranges that more readily command one's attention. By cultivating "natural" attention at the expense of the "cultural," however, we risk perpetuating the dualism which partly underwrites our problematic relationship with the more-than-human world. It's not clear, after all, that a month-long retreat to a secluded forest or mountain foothill would much improve one's attention to the quasi agency of waste plastic upon one's urban return. Nor would this retreat obviously heighten one's attention to other legitimate perspectives upon those consumer goods—including one's coltan-powered mobile telephone—which don't so readily resist instrumentalization.

Then perhaps, like Kohák, we should take the imperative to retreat to require "opting out" of urban experience. But this gives us a second problem. Since most of the aforementioned violent behaviors (e.g., unsustainable fossil fuel consumption, increases in consumer waste, or the rapid expansion of urban infrastructure) concern our relationships within the primarily "functional" world, fleeing to the woods won't much help us address them. Baldly put: by heeding Kohák's imperative, even in becoming attentive to one's *own* imposition of conceptual violence, one will be so far removed from more pressing intersubjective, "urban" concerns as to be almost entirely ineffectual in their disruption. Sojourns into nature in James's sense, then, can be, at most, part-time therapy for the task of attending to the attitudes, behaviors, and assumptions which epitomize the "functional" world we've constructed.

3.3.2 Hesitation

We need better means of cultivating attention *within* the "functional" world. Al-Saji aims to locate these via her rereading of Merleau-Ponty's late essay on painting, "Eye and Mind." The reason Al-Saji bases important political and philosophical considerations on ostensibly aesthetic or vocational ones is that the painter occupies a different epistemic role. The painter is distracted from the world in the purely functional sense because her primary concern is to see; "seeing is not subsumed to the ends of objectivating, goal-oriented action" (2009, 380). Painting may thus "take otherness as its motive rather than its object" (2009, 388). Cézanne's painting, Merleau-Ponty argues, "suspends [objectifying] habits and reveals the base of the inhuman nature upon which man has installed himself" (*EM*, 76). Heidegger echoes a related sentiment in his analysis of van Gogh's painting of a peasant's shoes. Van Gogh's painting reveals something meaningful about the peasant's world and, according to Al-Saji, about the social and political structures which shape it. As for the shoes, the peasant doesn't hesitate sufficiently to make the functional break; she "simply wears them" (Heidegger 1996, 160).

Unlike in Kohák's model, then, Merleau-Ponty's painter isn't "disconnected from the world, with all that this implies of distance and retreat. Rather, the painter's vision displays a different kind of attention to the world, to visibility and to life. It is an attention that seeks to remain within the proximity of seeing and seen, and tries to decenter, to question, this relation from within" (Al-Saji 2009, 380). For Al-Saji, even objectifying perception is already habitually attentive. But its attention centers upon the *objects* that emerge only against the background of the pervasive social and political influences hitherto explored. That background cannot, however, be registered within the naturalized discourse of extensional realism that, somewhat paradoxically, its commitments demand. Of course, the painter, like the peasant, is situated within a (somewhat similar) lifeworld, so her attention cannot result from disinterestedly bracketing her intentional commitments. Painterly attention, rather, "takes the form of hesitation; for it is hesitation that creates an opening in habits and makes them visible for themselves and within the world" (Al-Saji 2009, 380). In Al-Saji's account, then, painterly attention disrupts the unreflective utilitarian attitude of the "I can" from *within*.[20] Painterly hesitation is important, Al-Saji thinks, in allowing one to glimpse the diacritical structures generative of the "objects" perceived by painter and peasant alike.

Al-Saji's point is subtler than to suggest that we must become painters if we're to tackle patriarchy or climate change. She's really interested in *how* the painter becomes attentive to the diacritics of perception and not merely to the objects in which perception terminates. In Al-Saji's account, the painter's hesitation derives from her *affective* openness to the dimensions generative of objectifying vision. But this affective awareness isn't limited to Cézanne or van Gogh. As a human body-subject, Al-Saji argues, one is *always* affectively open to perception's background diacritics. Kelly's account neatly demonstrates this. Even within his somewhat objectivistic purview, one must be at least pretheoretically aware of some account of what lies behind the saloon's facade to perceive a substantive object at all. Without this sort of affective awareness, it's hard to see how one's perception of the film set could improve as Kelly, rightly, suggests.

Al-Saji's claim is nonetheless contentious in holding the scope of one's affective awareness to be much wider than Kelly implies. Al-Saji contends that we retain at least an affective awareness of some of the social and political dimensions generative of our propensity to perceive entities objectivistically. The background *abjection* of those dimensions, she thinks, is partly constitutive of the physiognomy of the objects registered by such a purview. Otherwise put: even if one isn't aware of dualistic logic *as* dualistic logic, for instance, the affective *weight* of its assumptions and commitments is already felt, albeit implicitly, in the revelation of a world furnished by determinate objects-in-themselves. Al-Saji suggests, then, that insofar as one may objectify the more-than-human world at all, one must be to some degree affectively aware that one is backgrounding—in a radically Plumwoodian sense—the aspects of that perception which cannot show up as objects-in-themselves.[21]

This is a strong claim and Al-Saji's arguments are highly technical, but her basic point can be illustrated with a couple of examples. Al-Saji (2009, 389) appeals to our pretheoretical awareness of the kind of intersubjectively sedimented structures (regarding the requisite ontological commitments to see races as natural kinds, for instance) that Sara Heinämaa and Linda Martín Alcoff highlight. We become reflectively aware of the affective weight they have in perception via the residual intentional "pull" they retain, even when disrupted (by reading Alcoff, for example). A more straightforward example concerns the abjection of other "unfamiliar" or "exotic" viewpoints which nevertheless form (what Judith Butler calls) objectifying vision's "constitutive outside" (Al-Saji 2009, 389). We see *this* kind of abjection clearly in Taiz's objection to floral intentionality. Seeing plants in Taiz's mechanistic manner requires *not* seeing them as intentional, because these purviews are "objectively" incompossible; they are irreconcilable according to mechanism's background commitments. Nonetheless, Taiz's abjection of "other" viewpoints is partly constitutive of his objectifying vision because seeing only objects requires *not* seeing intentions.

Nevertheless, this leaves aside why Al-Saji thinks we're *aware* of perception's diacritics before we reflectively uncover them. Al-Saji's reasoning is simple: if perception is irreducibly intentional *and* habitual, then some affective awareness of perceptual background is required to provide the means of establishing *different* embodied habits. I've suggested that one might *come to see* the world as reducible to "natural resources," or plants as biological machines, for example. However, to develop these differential operative habits, one's body must be, at some level, open to new intentional structures *and* able to retain (in a motor sense) the dimensions of antecedent structures as intentional sediment. Without some affective awareness—which a reflex arc account precludes—such habituation would be impossible. There must be, Al-Saji suggests, a certain "*indeterminacy* of habit" which "cannot be captured in the logic of objects" (2009, 382), but cannot be falsified by this contention, since affective indeterminacy is required to explain how one may acquire the very naturalistic viewpoint that denies it. I take Al-Saji to be suggesting, then, that, while dualistic and objectivistic prejudices may be ontologically inseparable from the perceptual worlds that most of us presently inhabit, the means to disrupt their diacritical structures exist within us in the form of habitual indeterminacy. Which is to say that the very means by which we acquire objectivistic sediment provide the means to develop better habits, because bodies must be fundamentally amenable to change.

Nonetheless, this discussion neglects *how* we're to develop sufficient habits of painterly attention to get a praxis of radical reflection off the ground. After all, the problem under discussion is that, as prereflective, any such affective awareness needn't be consciously available to us in the requisite sense to instigate radically reflective acts. But this is only partly true. There *are* occasions when one becomes more consciously aware (albeit only marginally) of phenomena which cannot be accommodated without a significant shift in one's background diacritics.[22] Al-Saji calls these disruptive events "inassimilable" (2009, 390), and Alcoff and Taiz provide neat examples of them. But various other inassimilable events have punctuated this book, including Bennett registering nonsubjective quasi agency, and Uexküll's frustrated attempt to wholly grasp tick temporality. In each case, the body-subject experiences a dissonance between the revealed phenomena and their own background commitments. Faced with such dissonance, one cannot help but hesitate and, thus, open some space for reflection, if (as Kohák claims) only momentarily.

The foregoing, I think, provides us with the requisite architecture for an account of how, despite the weight of objectivistic and dualistic sediment, one might instill a praxis of radical reflection that erodes one's entrenched violent tendencies. Let's put it all back together. Al-Saji has suggested that, as body-subjects, we're occasionally faced with inassimilable events which make us hesitate and bring to our attention some of our intentional commitments (i.e., those problematized by that event). This hesitation opens the requisite epistemic space for critical self-reflexivity. When faced with inassimilable events, one may reject inassimilable meanings as unreal (as Taiz does). However, where one admits their reality, but cannot reconcile them with one's current commitments, one can retain the status quo (a "pathological" response, but one which may be the norm under dualism). Alternatively, one can encourage marginal awareness of these new revelations to infuse one's world in much the same way that one's perceptual field reorganizes to admit gray squares, shipwrecks, or facades (Al-Saji 2009, 388). Through a radically reflective habitus of critique licensed by moments of hesitation, the latter option is possible, if difficult. I say that such a habitus is *possible* because one need only respond to inassimilable events by being radically reflective about the background commitments to which one is already sensitive. Doing so to some extent appears, to my mind, to be within the scope of autonomy as it presents itself phenomenologically. Given that embodied habits are indeterminate, where one hesitates with sufficient regularity, one may begin to disrupt one's objectifying habits and inculcate the radically reflective sort of humility that is conducive to addressing the residual violence secreted in one's (always provisional) view-from-everywhere.

Nonetheless, given Kohák's, Alcoff's, and Bennett's concerns about volitional self-reflexivity, shouldn't we be more pessimistic? The answer, I think, is no. Addressing the objectifying habits involved in reifying races as natural kinds, Alcoff herself says the following: "The multiple schemas operating in many if not most social spaces today would mitigate against an absolute determinism and thus pessimism. Perceptual practices are dynamic even when congealed into habit, and that dynamism can be activated by the existence of multiple forms of the gaze in various cultural productions and by the challenge of contradictory perceptions. To put it simply, people are capable of change. Merleau-Ponty's analysis helps to provide a more accurate understanding of where—that is, at what level of experience—change needs to occur" (2006, 189). Given our complex identities of oppression and privilege and the variety of conceptual and perceptual schemas in even "urban experience," we'll likely encounter inassimilable events on a fairly regular basis, which have the potential, where we aren't *entirely* pathological, to afford opportunities for radical reflection. I explore some in chapter 5. Furthermore, given the indeterminacy of habit, it's *possible* to become more open to such events where one, like the painter, makes a concerted effort to become more prone to hesitate. Where hesitation permits radical critical self-reflexivity and, particularly, where one uses hesitation to cultivate further openness to inassimilable events, this should afford optimism that the radically reflective ecophenomenologist might *better* perceive and engage with the more-than-human world as a matter of praxis.

3.3.3 Ecophenomenological Praxis

The reader might nonetheless wonder how exactly the aforementioned ontological, epistemological, and habitual shifts might translate into meaningful impact upon the environmental crisis. I've hitherto defined the praxis I prescribe as a kind of hyperreflexive practice of engaging with one's norms of interpretation so as not to violently impose metatheoretical assumptions and commitments upon the world. But I've said little specific about how the proponent of such a praxis might go about tackling oceanic plastic pollution, anthropogenic climate change, biodiversity loss, and so on.

There are, however, good reasons for this apparent oversight. One important reason concerns the modified characterization of environmental philosophy which I defended in chapter 1. There, I argued that ecophenomenological inquiry's value lies not in its capacity to straightforwardly tell us how to live but in its capacity to motivate the sort of habitual critical self-inspection which is anathema to unreflective objectification. Such a praxis may thereby help us, *ourselves*, to make better choices and inculcate more environmentally responsible habits. By expanding much further, I risk overstepping the mark and (somewhat ironically) undoing some of the good work I may have done in encouraging a radical humility about our epistemological, ontological, and axiological contentions, each of which is intimately tied to our intractably situated and embodied perspectives. To reprise an insight from the introduction, which I further justify by addressing interdisciplinary and intersubjective dialectics in chapter 5, advancing an ethic or policy framework is not the sort of task one can plausibly hope to achieve from one's armchair.

But this admission needn't prevent us from making any progress whatsoever in our attempt to disrupt the environmental crisis. I've recommended that this book be considered my contribution to the *groundwork* project of tackling the submerged mass of assumptions and commitments which underwrite it. To that end, disrupting the issue/solution schema (which might partly motivate the present objection) is a key ecophenomenological contribution. Moreover, I've suggested various useful tools (e.g., via the rejection of objective thought and other dualistic means of cleaving *us* from *them*; a more attentive view-from-everywhere; and my recommendation of habitual hesitation which permits greater attentiveness) with which committed and engaged theorists might begin to reconfigure their more specific commitments, behaviors, and conceptual and perceptual frameworks as a matter of habit. Each of these tools derives its value from its incompatibility with the uncritical, utilitarian, objectification of the more-than-human world, which, I've argued, is a common source of our failure to mitigate or overcome more specific anthropogenic ills. Likewise, by exploring shortcomings with the radically flattened ontologies proffered by ecofeminists, for instance, I've provided more concrete guidance on exactly what these conceptual and behavioral shifts *shouldn't* look like. I expand upon this point, and others pertinent to the development of an ecophenomenological praxis, in the coming chapters.

Nevertheless, in none of these chapters will one find the entailment of specific microlevel ecophenomenological principles by which anthropogenic climate change, for instance, should be overcome. This could be frustrating for the reader who shares my caution about ecophenomenology's scope but struggles to connect the aforementioned insights about the development of a critical social praxis with tackling more specific manifestations of anthropogenic violence. And, for obvious reasons, it would be a shortcoming of my project if it were to bear little upon how we encounter and relate to more specific aspects of the environmental crisis in at least an everyday sense.

However, the above caveats notwithstanding, it is *possible* to make (necessarily tentative and provisional) suggestions with reference to the argumentation hitherto advanced. Take the issue of climate change, for instance, for which our approach may have profound impacts at relatively high levels of decision making (e.g., by demanding that deliberative bodies be informed by diverse [e.g., nonscientific] disciplines and perspectives, without the standard division of labor, where "nonscientists" retain only a narrowly axiological role). Similar things might be said for town planning (e.g., by designing cities with "green spaces" positioned to frustrate the unreflective utilitarianism to which unsustainable fuel consumption may be linked), and more besides.

Related suggestions might be offered to situated individuals. We noted, for example, that taking Bennett's insights seriously should upset the impetus to produce and consume in our ordinary reckless ways, because the quasi agency retained by even our everyday "waste" renders it demonstrably dangerous and incapable of mastery by us. Earlier in the chapter, we aimed to do justice to these insights via their marginal inclusion in a nontotalizing view-from-everywhere, which should surely impact the specific habits we develop in response to anthropogenic climate change (e.g., by radically reconfiguring our "fuel" and material "needs"), oceanic pollution (e.g., by reducing or eliminating inessential plastic use and reducing reliance upon automated washing machines and other means by which microplastics and toxins reach oceans), and so on.

Likewise, the norm provided by the view-from-everywhere aims to better focus our attention upon the fact that even phenomena like foods are irreducible to their role in *our* various ordinary intentional narratives (e.g., as "mere meat," "composition of water, protein, carbohydrate, and nitrogen compounds," "something the children will eat," or "gustatorily pleasing combination of salt, acids, and fats"). Since our situated expressions are derivative from it, the view-from-everywhere is set up to encourage a heightened marginal awareness of other (often abjected) ways that entity might be differentially expressed (e.g., "[a portion of] something's body/offspring/corpse," "a drain on resources for the poorest communities," "a congealed mass of more-than-human labor and capitalistic economic imperatives linked to conspicuous consumption," or "someone's means of earning a living"). I've suggested that such marginal awareness should motivate a certain hesitant attentiveness through which one becomes more critically self-reflexive about whether one permissibly or justly backgrounds certain meanings via one's prioritization of expressions of others.[23]

In a similar vein, I've also suggested that one might become more sensitive to various axiological meanings within the world (e.g., the callousness *literally* manifested in our ordinary reduction of living, destructive, or toxic things to disposable use-objects). When contextualized within the architecture that Al-Saji's observations permit, our (albeit fallible) recognition of such meanings might encourage a more steadfast commitment to impactful and responsible action, perhaps sustained by a deep and real sense of injustice and/or heightened empathy. We should be reminded that our main target audience consists of those formally committed to reconfiguring their problematic relationships with the more-than-human world. I suspect, therefore, that sustained attention to these heterogenous meanings will motivate a reduction in (certain types of) animal product consumption linked to anthropogenic climate change via excessive fuel and water use and greenhouse gas emissions. Indeed, I suspect that sustaining the purview of humility demanded by the view-from-everywhere would probably lead to more sweeping changes in the choices of food we make, for example, and how and from where (or whom) we source it (situation permitting).

Of course, some of the above is relatively uncontentious guidance, even within mainstream scientific discourse. However, as Crutzen's earlier lament attests, it's a similarly uncontroversial observation that broad scientific consensus about what environmental "issues" demand hasn't thereby motivated the requisite attitudinal and behavioral changes to disrupt them at root.[24] This observation was pivotal in motivating my pursuit of an ecophenomenological *praxis* in the first place (i.e., partly as a means of translating our activist commitments into well-directed, habitual, action). Thus, since, even for the ecophenomenologist, "these thoughts are difficult to sustain—particularly so in the buzz and bustle of everyday life" (James 2009, 155), a further important question concerns how one might inculcate the requisite habitus of critique and mindset of attentiveness for sustained and well-directed action.

Earlier, following Al-Saji, I suggested that hesitation may come more easily through regular subjection to inassimilable events. If so, then where we seek to disrupt our own unreflective utilitarian or objectifying tendencies, it may be a good idea to restrict easy access to the consumer goods which perpetuate them (imaginative furniture reorganization is in order, at the very least). Likewise, rather than giving mainstream scientific discourse default de facto ontological priority, it would be strongly advisable to get into the habit of gaining our information about various multifaceted "issues" from diverse sources (e.g., news media with a variety of political leanings; media focusing on the perspectives of marginalized groups [or, better, direct dialogue therewith]; the testimony of people most directly affected; publications from multiple academic disciplines, and so on), and discussing them readily with differently situated Others.

There are probably much more radical and impactful changes to be made. While I reject Kohák's own radicalization of this thought, I take it to be an open question, for instance, the extent to which a *maximally* anticolonial, nonobjectivistic, and nondualistic habitus is ultimately compatible with the "Western" model of city living, and the communities and service provision typically cultivated therein. The devil, they say, is in the detail. But this sort of detail remains just out of reach. Insofar as it heeds its own warnings, such prescriptive guidance cannot be the goal of a specifically ecophenomenological project like mine, independent of substantive, ongoing, collaborative dialogue with theoretical and empirical work undertaken within various disciplines, and so on. The present work, then, is limited insofar as it aims to justify and outline a certain model of radically reflective praxis well positioned to expose the surreptitious conceptual violence behind our environmental crisis and open up broad routes of resistance. Nonetheless, as I hope to have demonstrated, subject to further work and extensive reflection upon our own roles as situated actors, it can connect with the important specifics of practice and policy-oriented guidance in the requisite manner to translate a more positive mindset into more directly impactful change.

3.4 CONCLUSIONS

We first investigated how an ecophenomenology informed by ecofeminist and new materialist insights might provide some means of better addressing the world on its own, rich terms. I suggested that radical reflection reveals scientific naturalism's reflective purification of "subject" from "object" to be badly made insofar as it underplays important continuities in the mind-like properties shared between human and nonhuman parties. By taking these subject-object crossovers more seriously, I argued, we may disrupt the assumed "objectivity" of the dualistic denigration of "mere nature." But perhaps more importantly, because epistemic "subjects" cannot be entirely divorced from their material situations, and because even the quasi-agentic actions of nonsubjects cannot be wholly predicted, we also undermine the project of mastery crystallized via a totalizing God's Eye view. Although the failure of the naturalist's transcendental project of purification means that the more-than-human world cannot be wholly possessed or controlled, I argued that this should give us no cause for regret as environmental philosophers. Rather, it should serve to motivate adherence to a more positive onto-epistemological norm in the form of the view-from-everywhere, which may better facilitate a reduction in violence through the richer range of meanings it admits. While one might nevertheless have concerns about the possibility of establishing the requisite shift in praxis to affect this reduction in violence, I argued that the indeterminacy of perceptual habits offers reason for optimism about the prospects for an ecophenomenology under which we may *better* perceive and relate to the more-than-human world as a matter of praxis. I ended by considering some tentative suggestions about how the praxis hitherto unpacked might connect, in impactful ways, with more practical and policy-oriented means of disrupting specific aspects of the present crisis situation, emphasizing some necessary limitations on our doing so.

CHAPTER FOUR

THE SPECTER OF CORRELATIONISM

Over the preceding chapters, I've explored some ways in which adopting a radically reflective ecophenomenological praxis might help us begin to disrupt some of the problematic assumptions and commitments which constitute the "submerged mass" of our crisis situation. Ecophenomenology, I've argued, is amenable to addressing more-than-human entities on their own, rich terms and is particularly well suited to fostering critical self-reflexivity about our habitual tendencies to overwrite more-than-human alterity. It's important to adopt a specifically *ecophenomenological* praxis, I've suggested, if we're to address the dualistic and objectivistic prejudices which imbue our default naturalism. If, as I've argued, these prejudices are key motors of the colonial attitudes which license our continued exploitation of the more-than-human world, then addressing and problematizing their residual sediment is an important part of addressing the environmental crisis. Thus—and, to reiterate, this is the main positive claim of my book—taking up the sort of praxis I promote surely has a key role to play in addressing the violence of our environmental crisis.

Over the remaining chapters, I shall explore some important potential challenges. In doing so, I intend to "flesh out" (excuse the pun) and further support the praxis I defend. Each of these challenges is intimately related to the "correlationist" charge we met in the introduction, and so it is there that we shall begin. Proponents of this charge hold that, because phenomenology is limited to concerns about how the world appears *for* a human subject, it is necessarily incompatible with the goals I have in mind. Since phenomenology precludes access to the modality of the in-itself, they claim, it cannot address nonhuman Others in *their* alterity. According to the speculative realists who level the charge, phenomenology maintains a fundamentally introverted focus which destines it to reveal nonhuman entities in a manner intimately related to their place in an (ontologically, and subsequently, ethically) anthropocentric schema. Thus, they claim, phenomenology is bound to perpetuate the colonial violence that I employ it to curtail.

In this chapter, I respond to the correlationist charge against Merleau-Pontian ecophenomenology via an argument with two steps. I first argue that if Merleau-Pontian phenomenologists court *something like* correlationism, this is because of a well-justified desire to retain the genuine alterity of other parties, serious engagement with which is better accommodated within *something like* correlationist philosophy. I then problematize the claim that Merleau-Pontian phenomenology is necessarily correlationist in any problematic sense. I hold that any serious version of the correlationist charge is misdirected because Merleau-Pontians deny the ontological primacy of the poles between which correlational "access" allegedly occurs. Ultimately, I argue that, by reinstating the centrality of the body to phenomenological investigation and, thus, licensing a *chiasmic* relationship between body-subject and world, a radically reflective ecophenomenology can nullify the problems supposedly introduced by correlationism even if, at some level of description, *something like* the charge may be true. Thus, I aim to show, there is no in-principle incompatibility between Merleau-Pontian ecophenomenology and the goals I set for it.

4.1 CORRELATIONISM

Speculative realism is a diverse movement united only by its formal rejection of correlationism and the desire to variously overcome it (Sparrow 2014, 115; Shaviro 2014, 65).[1] So, what is correlationism?

In a statement more or less echoed by all speculative realists, Quentin Meillassoux holds that "by 'correlation' we mean the idea according to which we only ever have access to the correlation between thinking and being, and never to either term considered apart from the other" (Meillassoux 2008, 5). Since the correlation is "unsurpassable . . . Correlationism consists in disqualifying the claim that it is possible to consider the realms of subjectivity and objectivity independently of one another" (2008, 5). Phenomenologists like Merleau-Ponty resist bifurcating epistemology and ontology because one's phenomenological opening is the only means of establishing any such ontology. However, as Merleau-Ponty argues, any phenomenology worth its salt recognizes that the body-subject actively structures the entire "concrete physiognomy" (*PP*, 57) of its experience. Phenomenology seems, therefore, only to be able to grasp the world as an intentional correlate *for* the body-subject, and thus, Meillassoux claims, phenomenologists *cannot* access any stipulative world outside the correlative schema.[2]

There are several strands to the criticism that correlationism (and therefore phenomenology) thereby *licenses* the sort of violence (e.g., technoscientific instrumentalism, large-scale environmental degradation, or exclusionary anthropocentric politics) I've been employing phenomenological tools to evade. I shall explore three in turn.

The Subjectivism Problem. If the bounds of the correlation are unbreachable, then "*nothing that is in the correlation can be used to adjudicate differing claims about what is outside the correlation*" (Richmond 2015, 400). Thus, for Meillassoux at least, by rejecting the "absolutism" of the thing-in-itself, "correlational reason thereby legitimates *all* those discourses that claim access to an absolute, *the only proviso being that nothing in these discourses resembles a rational justification of their validity*" (2008, 44–45). Since correlationism effectively reduces philosophical argumentation to fideistic belief, correlationism apparently results in a radical subjectivism, implicitly stacked in favor of the (human) subject, regarding how one should perceive, understand, and treat (nonhuman) objects. This isn't because any belief about a given entity can be put to a subjective court of appeal which somehow reasons to its ultimate justification, but because the fideism that correlationism allegedly results in "legitimates de jure every variety whatsoever of belief in an absolute, the best as well as the worst" (Meillassoux 2008, 46). Thus, although the correlationist affirms the fundamental (ontologically) anthropocentric asymmetry of the epistemic relationship between subject and object, under correlationism, any belief that a (human) subject might have about a (nonhuman) object appears to become equally legitimate simply by virtue of its being believed.

The Intersubjectivity Problem. The above is only part of the story. Correlationists allegedly have means *within* the correlation with which to adjudicate truth claims, but only through the sort of intersubjective consensus employed by Immanuel Kant, Edmund Husserl, and Donna J. Haraway.[3] One putative problem with such appeals is that, since the intersubjective community is effectively exclusively human, consensus entrenches a "species solipsism" (Meillassoux 2008, 50) whereby species norms usurp the primacy of any reality beyond it. Similar, or worse, effects are thereby licensed since species solipsism mistakes the human perspective for reality. Thus, we might plausibly expect anthropocentric instrumentalism to follow from correlationism in a manner similar to the one that concerned Heidegger about technology (i.e., by enframing things in terms of their use-value-for-us, albeit through a disinterested facade).

The Problem of Human Exceptionalism. Timothy Morton calls correlationism "anthropocentrism in philosophical form" (2011, 164) for some of the above reasons, but he also contends that the correlationist focus on acts of meaning constitution turns us away from concern for the world itself by imposing a mistaken human exceptionalism. This problem has two dimensions. In the first place, if objects exist only for subjects, and subjects are effectively ubiquitously human, then philosophy is drawn away from *concern* for "the real world" and toward *description* of "my world." This would prove disastrous in terms of a reduced push to address other entities on their own terms—terms which, speculative realists claim, cannot be "precisely the terms concomitant with the first-person phenomenological point of view" (Brassier 2007, 27). This shift also seems to speak against taking up the toil of environmental engagement since, as Simon P. James writes: "To think that mountain streams, humpback whales, and Californian redwoods need our care and attention, one must presumably see these things . . . as enjoying some kind of existence in their own right" (2007, 502).[4]

The second dimension of anthropocentric exceptionalism concerns turning away from the *real* things themselves, a move that apparently belies the ontological bifurcation of human subjects and (entities in) nature. Here, the bifurcation of subject and object imposed by correlationism allegedly contains, or facilitates, a *dualistic* hierarchy of kind, value, and the like. For Morton, this is largely because correlationism cleaves *human* subjects—who retain a constitutive monopoly on epistemological and ontological matters—from *natural* objects, which are plastic insofar as they cannot resist being violently subsumed under interested correlational terms. Thus, correlationism cannot obviously decenter the human perspective in the manner necessary for serious environmental concern.[5]

In these ways, philosophy since the "Kantian catastrophe" has allegedly been hopelessly and dangerously introverted because, even in acts of radical reflection, it cannot escape the "correlationist circle" (Meillassoux 2008, 5, 124). In this circle, entities cannot be addressed without the caveat "for us." For phenomenologists, the "for us" is explicitly tied to one's (human) intentional horizon. Thus, the argument goes, because of its paradigmatic correlationism, and regardless of any formal commitment to a radically reflective view-from-everywhere, ecophenomenology is unavoidably implicated in the imposition of colonial violence that I employ it to disrupt.

4.2 WHY REGARD FOR ALTERITY REQUIRES *SOMETHING LIKE* CORRELATIONISM

To respond to these objections, I will first discuss how the requirement not to overwrite nonhuman alterity, which infuses all three, may favor adopting *something like* a correlationist philosophy.

One plausible place to begin is Merleau-Ponty's analysis of the Other. As a matter of inescapable "perceptual faith," one recognizes that, as an alter ego, the other party opens onto a common world, which may nevertheless differ by significant degree from one's own (*VI*, 18, 141). There can, however, be no possibility of their perspective presenting itself to me without collapsing that alterity. If the Other were "given to me such as he is present to himself . . . the fusion of its ego with mine would cause its alterity to disappear" (Barbaras 2004, 128). Others are given to us fundamentally ambiguously: neither as pure positivity (object), nor—given the commonality of our world—as a pure lack (subject).[6]

If we cannot subsume the Other into ourselves by reducing it to an object, then we should presumably be attentive to the limits of our worlds and any "objects" therein. We've seen that addressing others in their alterity and evading an erroneous onto-epistemological imperialism requires first recognizing that "how the world is for me" cannot be unproblematically universalized. This is because, even as theorists, we're always both situated and irreducibly embodied.

A recap might be useful. Merleau-Ponty argues that epistemological and ontological claims rely upon the primacy of phenomenal perspective, in the context of which they (via the natural sciences and so on) derive their usefulness and truth-value. Importantly, Merleau-Ponty contends that one's embodiment here delineates the world's physiognomy, since our expectations of reality are tied to our bodily capacities as our medium of having a world in the first place. "A thing," Merleau-Ponty argues, is "not actually given in perception, it is internally taken up by us, reconstituted and experienced by us in so far as it is bound up with a world, the basic structures of which we carry with us" (*PP*, 326). One's primordial relationship with(in) the world is, therefore, operatively intentional; one's embodied expectations codetermine the world's meanings via one's grasp on it. Merleau-Ponty also argues, however, that perception is subject to sediment via the intentional arc. Here "knowledge" is sedimented into habits which influence the body schema at the prereflective level. What all of this means is that reality is always expressed from within a situation which one cannot plausibly transcend because, even if one were (somehow) able to bracket all historico-cultural sediment and return to a presedimentary situation, this would still be undertaken *as* a concrete human body, the grounding perceptual norms of which won't straightforwardly map onto all biotic entities, let alone allow one to exhaustively express the perceiver-independent reality of any putative abiota-in-itself.

I repeat these details to emphasize that recognition of one's contribution to the prethematic reality which underwrites our ontological and epistemological claims is given central prominence in Merleau-Pontian phenomenology. One implication is that philosophy, for Merleau-Ponty, is properly characterized *as* radical reflection: the perpetual process of attempting to slacken one's intentional threads to better appreciate the basic intentional structures that one coconstructs with the world. But, as we've seen, Merleau-Ponty is also expressly concerned with the chiasmic encroachments that radical reflection might reveal with other body-subjects and, consequently, with overlaps *and* inexhaustible differences in the meaningful ways the world might be expressed. If one's differential embodiment is as epistemologically pivotal as Merleau-Ponty claims, then one task for radical reflection *must* be to maintain a critical focus on uncovering where one might (implicitly) overwrite alterity via the imposition of situated (and sedimented) ways of perceiving as universal truth. Merleau-Pontian ecophenomenology may thereby take entirely seriously the thought that expressive truth rules out the plausibility of exhaustive accounts of (particular aspects of) reality by taking to heart Emmanuel Levinas's thought that objectification involves subsuming all truths into mine; it is "the *reduction of the other to the same*" (Levinas 1987a, 48). Merleau-Pontian ecophenomenology is, after all, principally defined by its hostility to the totalizing ontological and epistemological objectification that Levinas criticizes.[7] This enables Merleau-Pontian ecophenomenologists to set out their *anticolonial* stall.

Given that "man is in the world, and only in the world does he know himself" (*PP*, x–xi), there can be no *coincidental* return to "the things themselves" in Ray Brassier's sense. This *seems* to make Merleau-Pontian ecophenomenology correlationist. However, if *something like* correlationism is the price to pay for an intrinsic focus on addressing one's own imposition of onto-epistemological violence, then perhaps Merleau-Pontian ecophenomenologists should cheerfully accept the charge (or something like it).

But this may be too quick. Speculative realists argue that *they* are suitably reflexive about such violence, whilst also maintaining the metaphysical realism apparently essential to avoid disastrous environmental implications. After all, speculative realists do not claim unfettered access to reality-in-itself. Rather, they claim, reality can be (speculatively) known *in part*. It seems, however, that the brand of realism they employ ultimately undermines their capacity to maintain sufficient reflexivity.

4.2.1 CORRELATIONISM, REALISM, AND IDEALISM

One underlying problem derives from speculative realists mistaking phenomenological "correlationism" for an anthropocentric absolutizing of the subject pole (in the problematic ontological sense aligned with human chauvinism) and, therefore, something to be avoided at all costs. In the speculative realist literature (e.g., Meillassoux 2008, 18; Sparrow 2014, 26), a binarism is often set up between idealism and realism whereby denying *epistemic* access to the thing-in-itself is considered enough to commit one to a full-blown *metaphysical* idealism. Although I cannot discuss it in depth here, this claim is misleading, even when leveled at the arch-correlationist: Kant.[8] To his credit, however, Graham Harman attempts to address this conflation of epistemology with metaphysics head-on, so we will evaluate his argument more fully.

Harman argues that correlationists are committed to a de facto metaphysical idealism because correlationism precludes access to the world-in-itself and, thus, entails that the world is *treated* in an idealistic manner by the correlationist. That is, under correlationism one absolutizes the subject pole as a matter of praxis and so correlationism legitimates the violent behaviors mentioned above. If one lacks epistemic access to the glacier in-itself, the argument goes, the glacier is effectively reduced to the ways it appears to perspectives *like mine*. Harman therefore proposes a "litmus test" for correlationism: "Of any philosophy we encounter, it can be asked whether it has anything to tell us about the impact of inanimate objects upon one another, apart from any human awareness of this fact" (2005, 42). Answering "no" to this question, Harman thinks, "condemn[s] philosophy to operate only as a reflexive meta-critique of the conditions of knowledge" (2005, 42).

We should have reservations about much of Harman's argument, but I want to focus on one response available to the ecophenomenologist. It might be claimed that Merleau-Pontian ecophenomenology can *pass* Harman's litmus test and, thus, escape the charge of de facto idealism. It may do so because, as we've seen, Merleau-Ponty thinks we only perceive whole objects (marbles, film sets, rainforests, and so on) because perspectives *licensed by other entities themselves* partly constitute our perceptions of them. Thus, one might claim, Merleau-Pontian ecophenomenology can tell us *something* about interobjective relations without reducing them *to* one's awareness of those relations in the requisite idealist manner.

To clarify: accepting the partiality of situated perspective is the starting point of object-oriented ontology. Harman cannot, therefore, think that the admission that one's knowledge of interobjective encounters is always situated is enough to render a philosophy idealist without undermining the realism of his own project. Neither can Harman claim that idealism is entailed by the mere fallibility of any such knowledge claims. Like Meillassoux, Harman admits the fallibility of speculative knowledge claims. Regarding other entities, there needs to be a stronger epistemic "reduction to," or *metaphysical* "dependence upon," human awareness to entail Harman's de facto idealism. But Merleau-Ponty doesn't invoke either of these.

Nevertheless, Harman cites the passage in question as an example of Merleau-Ponty's idealism, calling it "a metaphysics of relations" (2005, 50). Harman (2005, 51) argues that if something is reduced to a view-from-everywhere which leaves "nothing hidden" (as Merleau-Ponty claims), then Merleau-Ponty *is* effectively advancing an idealist metaphysics. This is because, Harman contends, the entities in question *are* ultimately nothing more than the multiple human perspectives in which they feature. After all, Merleau-Ponty doesn't (and shouldn't) engage in the panpsychist or animist anthropomorphism under which glaciers or rocks straightforwardly perceive one another in the absence of intentional body-subjects.[9] A metaphysics of relations, for Harman, is tantamount to idealism and is purportedly exemplified by Merleau-Ponty's consistent reduction of the in-itself to an "*in-itself-for-us*" (*PP*, 375). Harman even accuses Merleau-Ponty of a pseudo-Kantian intellectualism under which reality is "a vast homogenous totality until humans burst onto the scene" (2005, 52).[10] If Harman's critique holds, then Merleau-Pontian ecophenomenology may absolutize the human subject in a suspiciously anthropocentric manner where its ontological predilections align with a problematic kind of ethical instrumentalism.

Harman, however, misunderstands what Merleau-Ponty means by the "in-itself-for-us."[11] For Merleau-Ponty, the *inexhaustible* contribution of the world with(in) which one communicates is an essential codeterminant of reality that is irreducible to Harman's idealistic characterization. Although the world shows up only in particular expressions as integrated, diacritically interrelated wholes, for Merleau-Ponty, the world retains what James calls a "brute presence," irreducible to acts of perception, which is the very reason why our intersubjective world is unavoidably *common* (James 2007, 507; *PP*, 320–24). Thus, although the view-from-everywhere provides normative onto-epistemological guidance, entities aren't thereby reduced to the sum of all "subjective" perspectives, because they coexpress themselves *in* or *through* those perspectives. I might be unable to grasp a preexpressive, autonomous, reality, but reality doesn't thereby become *mine* in any accurate sense. There is, as we've seen, a meaningful distinction to be made whereby "I cannot say that *I* see the blue of the sky in the same sense in which I say that I understand a book or again in which I decide to devote my life to mathematics."[12] One cannot *co*mmunicate alone.

However, Harman or Meillassoux might respond that this still doesn't satisfy the more fundamental objection that phenomenology remains idealistic because of its incompatibility with epistemic access to objects or properties *in-themselves*. Meillassoux would contend that, since Merleau-Ponty concedes that the view-from-everywhere can be only a "presumptive synthesis" (*PP*, 90), rather than being straightforwardly ontologically manifest, Merleau-Pontian phenomenologists remain unable to say anything about the perceiver-independent absolute. Other promising examples Kelly cites, such as when "the lighting directs my gaze and causes me to see the object, so that in a sense it *knows* and *sees* the object," fare little better because, for all the object's extrahuman impetus, if the "view from nowhere . . . is a contradiction," then the object cannot be understood entirely independently of the gaze in which it manifests itself (*PP*, 67, 310).

While these contentions about phenomenology's inability to access epistemically-purified things-in-themselves are true as far as they go, they serve to highlight the fundamental problem with the charge of idealism: speculative realists take the subject/object cleavage to be ontologically primitive in a manner denied by Merleau-Pontian ecophenomenologists.[13] Thus, where one cannot access reality-in-itself ("objective" reality), one is apparently committed to a "subjective" prison, in which reality is subsumed into one's *own,* narrow, perspective. But carnal phenomenologies employing a doctrine of expressive truth simply don't conform to this rigid subject/object schema (with its dualistic trappings) other than as a reflective abstraction. Moreover, the main reason that Merleau-Ponty rejects objective thought in the first place is that, by reifying a purified realm of objects, we lose sight of our important contributions to their basic intentional physiognomy. I've suggested that it is these contributions that radical reflection must explore if we're to *prevent* the erroneous and dangerous imposition of ontological and epistemological terms.

In fact, the acquired subject/object schema that underpins speculative realist analyses plausibly derives from the primary ambiguity experienced by body-subjects: an ambiguity necessitated by taking one's embedded implication in perspectival Gestalts seriously. But this ambiguity doesn't make the "subject" the sole motor of truth, and this is where the charge of idealism breaks down.[14] Merleau-Pontian ecophenomenologists can, and do, take seriously the contributions of other entities on their own terms. They simply recognize that doing so *also* requires radical critical self-reflexivity to the limitations of one's situated perspective. Nonetheless, the important fact remains that, while narrowly *self*-reflexive critique retains a crucial role in Merleau-Pontian phenomenology, it cannot be *the* task of philosophy because its focus is too narrow. It would recognize only one strand of radical reflection (critique) at the cost of the other (openness to the world).

If my arguments hold, then Merleau-Pontian phenomenology needn't be guilty of the idealism which (along with dogmatic metaphysicians' failure to recognize their own value-laden, ontologically anthropocentric, heritage) motivates a speculative turn. If so, then speculative realists also appear to be misled about the limitations of phenomenology's scope. Their analyses do, however, remain helpful in foregrounding the principles upon which speculative realism is built and which will ultimately speak against *its* ability to address more-than-human entities in their alterity.

4.2.2 Tensions between Anticorrelationist Realism and Alterity

In addressing the modality of the in-itself, speculative realism's focus subtly, but importantly, shifts away from addressing one's own limitations and commitments, and toward stipulative transcendent means of access which nullify one's subjective contribution. Even putting aside Merleau-Ponty's arguments for the implausibility of doing so, this shift of focus turns one's reflective attention *away* from the -centric overwriting of alterity: one of the very issues that was to be addressed by speculative realism. Steven Shaviro, for example, blames Kant's alleged "anthropocentrism" on his critical self-reflexivity which, Shaviro contends, should be "dislodged" because it is "too inward-looking" (2014, 72). What tends to happen, instead, is that speculative realism is drawn toward reflective abstractions. Abstracting in this manner is problematic, as we've seen, for two reasons: because the risk of reflective distortion increases the further one abstracts from grounding perceptual expressions, and because such abstractions are *more* prone to conceal the contribution of the situated body-subject. Here are a few examples from the speculative realist literature to illustrate.

To avoid relying upon a phenomenological opening, both Meillassoux and Brassier reduce the in-itself to inert matter. They each do so because they think that disrupting the correlationist circle requires teasing apart subject/thought from object/being. By purifying the in-itself from reliance upon (human) thought or subjectivity, they hope to speculatively access things-in-themselves on their own terms. For Meillassoux, radically decentering thought requires resuscitating a neo-Cartesian account of primary qualities under which only "*those aspects of the object that can be formulated in mathematical terms can be meaningfully conceived as properties of the object in itself*" (2008, 3, 115). For Brassier, a significant "gain in intelligibility" comes through the realization that being-in-itself is characterized by "the extinction of meaning . . . the cancellation of sense, purpose and possibility" (2007, 238).

As Plumwood argues, however, these sorts of schematization belie a truncated reversal of the Cartesian schema: a misleadingly anthropocentric metaphysics—minus the mind—under which nonhuman "objects" are implicitly instrumentalized. By leaving the subject/object dualism's dogmatic terms unchallenged, truncated reversals retain the hierarchical assumption that "objects" differ in kind from (human) "subjects" in a manner that effectively legitimates viewing the former as a resource. Simply jettisoning the subject pole, as Meillassoux and Brassier do, won't disrupt the residual political agenda because only an impoverished conception of the world is left behind (Plumwood 1993, 47).

But Meillassoux's problems don't end there, because his "reversal" isn't even consistently "truncated." Meillassoux's problematic human exceptionalism is best seen in his account of the "brute emergence" of subjectivity with humanity, which would make human thought radically discontinuous with the rest of the world (Harman 2011a, 59; Shaviro 2014, 75). Brassier, however, fares little better. His exceptionalism comes with the nihilism he equates with the absence of *human* valuers. Both cleave mind/culture/subject from body/nature/object along problematically dualistic lines that should trouble Morton as much as Plumwood.

Shaviro attempts to address these issues by rehabilitating a "paradoxical," Whiteheadian mode of reflective speculation, which, he claims, is our only means of encountering "the world without us obliquely" (2014, 66–67). However, Shaviro curtails the subject/object bifurcation too forcefully and *collapses* the distinction into a panpsychism at the level of the in-itself, thereby subsuming nonhuman entities under human categories, notably *universal* sentience. Although we've seen Merleau-Ponty argue that intentionality extends well into the biotic community, this only problematizes the body/mind binary. Significantly, it means that, contrary to Harman's (2005, 173) assertion, Merleau-Pontian Others needn't be human.[15] However, taking seriously the possible heterogeneity of mindedness and its relationship to our concrete modes of embodiment, Merleau-Ponty's admission certainly needn't entail Shaviro's contention that "intentionality is . . . an ontological feature of objects in general" (Harman, cited in Shaviro 2014, 80).[16]

To clarify: while I applaud Shaviro's rejection of Meillassoux's assumption that "thought, value, and experience are essentially, or exclusively, human to begin with" (2014, 91), this revelation doesn't lead Shaviro to problematize the Cartesian sediment that underpins Meillassoux's account. Shaviro ultimately admits nonhuman entities to the sphere of mindedness, not by disrupting the subject/object dualism's hierarchical and atomistic terms, but by anthropomorphically extending them without reserve. Thus, to the extent that Shaviro may address nonhuman Others, their subjectivity cannot differ from his in any radical manner. Shaviro says as much when he claims that "the bat's inner experience is inaccessible to me, but this is so in much the same way . . . that any other person's inner experience is accessible to me" (2014, 92). By making *all* entities "autonomous centers" (2014, 89) of life with remarkably humanoid subjectivities, Shaviro appears to be guilty of what Plumwood calls "incorporation": the colonial act of defining the underside in a manner relative to the master identity—here: roughly, a Cartesian subject—rather than emancipating it in *its* alterity.[17]

The "object-oriented ontology" strand of speculative realism is tainted by similar problems. Object-oriented ontologists make an important concession to alterity in recognizing that, as one situated entity among others, one cannot exhaust the reality of the in-itself. Object-oriented ontologists hold that differential objects-in-themselves (e.g., human beings, pins, rainforests, axolotls, or neutrinos) "enter into relations," yet "their allies cannot mine their ores" because the objects *themselves* remain hidden as they "withdraw" (Harman 2009, 132). Interobjective communication is effected by decoupled properties which "metaphorize" one another, "distorting [other objects], caricaturing them, bringing them into play only partially" (Harman 2005, 172). Following Harman, Ian Bogost explains that metaphorism captures "some aspects of something else at the cost of other aspects," based upon the "internal properties" of the object doing the caricaturing (2012, 66).

The problems here are twofold. First, alterity *is* retained via object withdrawal, but only in something akin to Levinas's sense of a "nonreciprocal relationship" in which the withdrawing Other stands impenetrably apart.[18] Like Harman (2005, 103), Levinas thinks that where the Other is made present to consciousness she is, at best, encountered as an *alter ego* derivative from oneself. *Absolute* alterity, Levinas claims, is therefore precluded by phenomenologies of a Merleau-Pontian ilk. In a similar vein, object-oriented ontologists charge ecophenomenology with correlationism (and idealism), because it allegedly subsumes nonhuman entities into a humanoid intentional horizon.[19] Object-oriented ontologists appear beholden to something like an *absolute* conception of alterity because their analyses depend on the binary yes/no question of whether the *object*-in-itself is made entirely manifest to another. Since all interobjective encounters allegedly take place in the modality of the in-itself but differ significantly—rainwater, for instance, can wash away spilt vinegar, yet never *itself* interacts with its pungent taste—object-oriented ontologists require objects to retain an excess unexhausted in any interobjective encounter.[20] Thus, what is encountered, object-oriented ontologists hold, can never be the object itself.

Merleau-Ponty, however, would think that this absolute, impenetrable, conception of alterity *precludes* serious engagement with other entities. By "fetishizing" alterity, Jack Reynolds (2004, 317) argues, one undermines the very possibility of (even partial) epistemic encroachment, solipsistically hyperseparating parties such that addressing them on their own terms becomes impossible.[21] But the problem runs deeper. Like Simone de Beauvoir, Sara Heinämaa (1999, 125–26) argues that the assumption of this kind of alterity has unwelcome political reverberations since it licenses the self/other *dualism* and the violence subsequently legitimated toward those marked as "other." In short: while the deployment of absolute alterity may help limit anthropocentric hubris in our ontological and epistemological contentions, it doesn't really allow us to address nonhuman Others at all. Object withdrawal, therefore, does little to foster engagement with more-than-human entities in *their* alterity, which was the primary requirement of the issues surveyed in section 4.1.[22]

Genuine engagement with more-than-human entities seems, instead, to require the right balance between addressing their "openness" to us and maintaining attention to their "occultation" (*VI*, 28). Doing justice to alterity here requires two things: first, that more-than-human entities must be addressed on their own terms, irreducible to one's own; second, recognizing that, *as other*, other entities will never exhaustively deliver their truths. Both requirements are compatible with the doctrine of expressive truth central to Merleau-Pontian ecophenomenology. This is because, for Merleau-Ponty, an entity is only ever *partly* present to a body-subject via its brute contribution to a given expression.

The second problem with the object-oriented account is that alterity *isn't* given sufficient due with respect to the metaphorizing relations between objects. By the object-oriented ontologists' own lights, they must either accept that one cannot speak meaningfully of *nonhuman* inter-objective communications because one always metaphorizes Others (so I metaphorize [i.e., anthropomorphize] relationships between rainwater and vinegar, or between orangutan and rainforest), or else one must somehow transcend one's situation to describe such communication accurately. Attempts to do the latter are inevitably couched in markedly human terms, either via reductive scientific terms (usually about quanta) which are presented, bizarrely, as if universal (e.g., Morton 2011; Bogost 2012), or suspiciously anthropomorphic descriptions under which objects' properties interact in exactly the same terms as human (or humanoid) experience. Morton says, for example, that "what spoons do when they scoop up soup is not very different to what I do when I talk about spoons . . . because intelligence and being alive are aesthetic appearances for some other phenomenon, including the object in question" (2012, 215). In Harman's famous example, the properties with which fire and cotton metaphorize one another, and those which withdraw (e.g., the cotton's "aroma" or the fire's "foreboding sparkle"), are freely available to *us* humans (2005, 170). There's little residual unfamiliarity. Nor does there seem to be much chance of our surprise at the revelations the world itself throws up. Both observations sit discordantly with object-oriented ontology's ability to engage with a genuine regard for alterity, or a real decentering of the human on more-than-human terms.[23]

In order to make these more audacious metaphysical claims, there appears to be an ambiguity in what is meant by the excess that withdrawing objects retain. For object-oriented ontologists, objects are irreducible to a metaphysics of relations; objects never manifest *themselves*. But this claim alone tells us nothing about *how* something's alterity is retained through that excess. Doing justice to alterity, I've suggested, involves appreciating important continuities and differences. According to the Merleau-Pontian line I've been pursuing, an entity's excess must include (the power to instantiate) some properties or qualities which are straightforwardly unavailable to situated human "objects" in virtue of the kinds of entity they are. This thought also resonates with some of the object-oriented ontologists' official commitments. Harman, Morton, and Bogost argue, for instance, that one never directly addresses the thing-*in-itself* for the very same reason as Merleau-Ponty: doing so would mistakenly reduce it to a situated (anthropocentric) perspective. However, in examples like those in the paragraph above, excess seems to mean something less radical, something more like "retains properties not *currently* shared with the other object." Morton (2011, 165) gestures toward this conservative interpretation when he explains object withdrawal via the analogy of a coin which cannot show both sides simultaneously but retains no substantive properties that are unavailable to us in-principle as human perceivers. One might never see the coin's other side *as* its other side (i.e., as its excess), but one *can* exhaustively describe its features after flipping it over.[24]

This conservative interpretation of excess would explain why Harman and Morton feel justified in making the determinate claims they do about the metaphorizing relationships between objects. If object-oriented ontologists mean to propose this conservative meaning, however, they seem to be committed to the sort of naturalism that they distance themselves from on the well-justified grounds that it is incompatible with substantive nonhuman alterity. Thus, as Stacy Alaimo puts it in her paper on Bogost, although object-oriented ontologists put "forth an ostensibly posthumanist ontology, I would contend that [they retain] a humanist and masculinist sense of a disembodied subject" (2014, 15).

Object-oriented ontologists recognize some of these dangers. They evidently *intend* excess to be understood in the more radical sense where objects really do withdraw from one another in some substantive manner. Doing so, however, means that their (situated) claims about excess or withdrawal are incompatible with the desituated and totalizing metaphysics they advance. Moreover, in making specific claims about the "true" nature of other "objects" (including those we're attempting to liberate from uncritical instrumentalization: humpback whales, coral, crude oil, and so on), they are returned to the de facto idealism entailed by an absolutist account of the self/other relationship. Interestingly, if object-oriented ontology did conform to the radical interpretation of excess, it would *fail* Harman's litmus test for correlationism because it couldn't describe interobjective communications (e.g., between nonhumans) independently of a situated perspective that "distorts" or "caricatures" them.[25] Thus, Dan Zahavi (2016, 296) is surely right to argue that, through a radical doctrine of object withdrawal, Harman and his colleagues adopt the very metaphysical antirealism that Harman equates with de facto idealism.

As I see it, the underlying problem here is that speculative realism's stipulation that reality cannot be accessed from within the correlation introduces a false dilemma: to absolutize the object pole (Meillassoux, Brassier, object-oriented ontology) or to collapse the subject/object distinction in the modality of the in-itself (Shaviro). Since both options are plausibly derivative from the phenomenological reality chiasmically coexpressed between world and body-subject, both erode alterity through egomorphic or anthropomorphic imposition. Why? Because the speculative realist requirement for unmediated access to the in-itself requires an objectivistic reduction to the same, yet "the same" is always gleaned from a situated and partial (if not strictly singular) grasp of the world. Moreover, by shifting focus away from critical self-reflexivity, speculative realists appear to be in a worse position than their ecophenomenological rivals to diagnose and attend to such problems. If I'm right about this, then, in lieu of recourse to the sort of radical reflection available to Merleau-Pontians, speculative realism is in a worse position than Merleau-Pontian ecophenomenology to engage with more-than-human entities *in their alterity.*

4.3 IS MERLEAU-PONTIAN PHENOMENOLOGY CORRELATIONIST?

One important consideration here surrounds any putative "realism" adopted. As I noted, Merleau-Ponty *does* retain inescapable categories of truth and reality, but these are grounded in perceptual coexpression rather than transcendent acquaintance with things-in-themselves (*PP*, xvi). As touched upon in the previous chapter, the mere possibility that a meaningful expression might be subsequently shattered and replaced (such as when we realize a "rock" is really a lump of driftwood, or that a "saloon" is a facade) doesn't "amount to sceptical doubt" about the world "because, in short, doubt must be *lived*" (James 2009, 48; *VI*, 40). Thus, the praxis of affirmation and seriousness with which the Merleau-Pontian phenomenologist addresses the *real* world mirrors that of avowed metaphysical realists.

Another salient aspect of the Merleau-Pontian account is that, since one *is* one's body, one is implicated in the rest of the world via one's carnality. Thus, the relationships which underpin expressive dialogue are, as James notes, significantly more "intimate" (2007, 512) than those between subject and object. As a body-subject, one has direct epistemic contact with the world in a manner not true of Kant (or, perhaps, Husserl), because one isn't ontologically bifurcated from it as Cartesian subject/object dichotomies apparently require. Speculative realists, however, are unsuited to make sense of this epistemic relationship since their metaphor of "access" requires the problematic reification of ontologically distinct poles from the outset.[26]

Furthermore, through one's embodiment, one may share significant ontological continuities with other animals and, albeit to a lesser extent, other (a)biota. Kelly Burns, for example, notes that, for Merleau-Ponty, "the similarities in the ways that our bodies operate in the world lead to similarities in consciousness, which is not an isolated phenomenon, but a common experience" (2008, 110). One shares significant bodily continuities with certain other mammals, for instance, so it's plausible that at some ideal, presedimentary level, there are common (if gradated and heterogeneous) nondiscursive grounds of experience that testify to a somewhat shared, *real* world.[27] Once one dispenses with the requirement for a rigid subject/object or self/other binary, engagement with other entities *in their alterity* becomes feasible on these grounds.

We may go on. Elizabeth Grosz (2005) and Brian Massumi (2015) emphasize human-nonhuman connaturality through evolutionary continuity. Massumi draws attention to the evolutionary "supernormality" of herring gull behavior which also infuses human creativity.[28] Merleau-Ponty (*SB*) makes a related point by appealing to the chiasmic Gestalts of matter, life, and mind, under which something novel (i.e., life or mind) emerges from its predecessor.[29] Merleau-Ponty argues that, as embodied, "higher" Gestalts rely upon ever-present but subordinated Gestalts that condition them. Subordinated levels are associated with more rudimentary ways of encountering the world. Nevertheless, as Ted Toadvine contends, through the chiasmic contribution of subordinated levels to our grasp, we share partial epistemological and ontological crossover via "lateral kinship" with other entities, since we're connatural with them (2014, 274–75; *VI*, 207–8).[30]

We've also seen Jane Bennett argue that the quasi-agentic contributions of "other" actants infuse human perception. If, at some level of description, as Bennett claims, we're (interrelated) assemblages like pretty much everything else—abiota included—we have further reason to believe that there exist *some* such nondiscursive grounds. Examples abound.

The preceding discussion suggests that it's plausible, via the chiasmic interrelationship of subject and (other entities in the) world that one *is* as a body-subject, that one may be well-positioned to express, to some degree, how things are for (an aspect of) the world *itself, without violating the alterity requirement*.[31] This is, in short, because the charge of "anthropocentrism" fails to recognize that, as a body-subject, one isn't *exclusively*, nor *entirely*, "human." To say otherwise would be to bifurcate nature and mind in a manner spoken against by speculative realists and which forms part of Morton's argument *against* correlationism. Thus, we may respond to Meillassoux's two objections to what phenomenological "correlationism" entails. On the subjectivism problem, Meillassoux is mistaken because nondiscursive grounds give inescapable bases on which to assess truth claims. On the intersubjectivity problem, Meillassoux construes intersubjectivity too narrowly. By bifurcating mind and body along Cartesian lines, Meillassoux fails to recognize that one's body implicates one *in* the world addressed, and Others therein, in an important, if partial, manner.

We can see, therefore, contra Morton's charge, that Merleau-Pontian ecophenomenology retains a robust kind of realism. While phenomenologists may be realists in the sense of engaging with a concrete and inassimilable reality accessed noninferentially, however, they are unable and unwilling to commit to the definition of realism (i.e., by addressing metaphysically distinct things-in-themselves) required by speculative realists. If the dispute about whether Merleau-Pontian phenomenology is correlationist turns *entirely* on a terminological point about the definition of "realism," however, then the charge seems to me empty, or not very serious.

There is, however, a remaining strand of the correlationist charge we have hitherto glossed over. Given that Merleau-Pontian ecophenomenologists refuse to exhaustively assimilate "how things are for the world" into "how things are for me," this leaves them open to Meillassoux's *modified* correlationist charge: "We can't *know* what the reality of the object in itself is because we can't distinguish between properties which are supposed to belong to the object and properties belonging to the subjective access to the object" (Meillassoux et al. 2007, 409; emphasis mine).[32] This claim, in part, I accept. It's exactly this sort of issue that necessitates a radical habitus of critique in the first place. However, since Merleau-Pontian ecophenomenologists deny the basic ontological presuppositions which underpin the charge, this admission still doesn't make Merleau-Pontian ecophenomenology correlationist in any substantive or problematic sense. Moreover, given that the radically reflective framework of Merleau-Pontian ecophenomenology is partly driven by a desire to better understand the more-than-human world on its own terms, the critically self-reflexive outlook it necessitates seems to be at odds with the detrimental outcomes supposedly caused by correlationism. Furthermore, to the extent that it is *speculative* rather than dogmatic, given my arguments in section 4.2, neither can speculative realism plausibly escape Meillassoux's modified charge. By drawing our attention away from critical self-reflexivity and overstating the surety of our distinctions between "objective" and "subjective" properties, speculative realists appear to be at greater risk of instigating and perpetuating such outcomes.

4.4 CONCLUSIONS

I've suggested that engaging with more-than-human alterity provides compelling motivation for *something like* correlationism. However, since Merleau-Ponty outright denies the ontological primacy of the "subject" and "object" poles upon which the correlationist charge is built, doing so needn't make Merleau-Pontian ecophenomenology correlationist. There are two reasons why Merleau-Pontian ecophenomenology might nevertheless be understood to be *something like* a correlationist philosophy. The first is that it cannot satisfy the speculative realist's objectivistic stipulations about what metaphysical realism consists in. The second is it cannot straightforwardly cleave contribution of "object" from "subject" in the requisite manner to ascribe properties to things-*in-themselves*. But neither should be particularly troubling. I've argued that the former stipulation is unimportant or even harmful because of the violent and dualistic assumptions it conceals. As for the latter, I've argued that the Merleau-Pontian ecophenomenologist's refusal to cleave subject from object is strength inasmuch as it disrupts the transcendental fiction that limits or discourages critical self-reflexivity about our tendencies to overwrite more-than-human alterity. I contend, therefore, that it's implausible to claim that Merleau-Pontian ecophenomenology is more likely to precipitate colonial anthropogenic violence than its markedly anticorrelationist rivals. The "correlationist charge" itself, therefore, provides no reason to think that a Merleau-Pontian ecophenomenological praxis is incompatible with the goals I set for it.

CHAPTER FIVE

ANDROCENTRISM, NONDISCURSIVE GROUNDS, AND THE HYPERDIALECTIC

I begin this chapter by considering a narrower version of the correlationist charge raised primarily by feminist philosophers. This charge holds that, while phenomenologists face no in-principle impossibility in addressing the more-than-human world on its own terms, phenomenology is, in practice, hostile to alterity. According to this narrower critique, phenomenology is a particularly dangerous form of something like correlationism because of its underpinning *androcentric* norms. Thus, proponents claim, rather than doing justice to the primordial or anonymous nature of one's being-in-the-world, phenomenologists—particularly Merleau-Ponty—universalize situated master perspectives in a manner that speaks against sensitivity to the ecofeminist concerns explored in chapter 2.

This charge is more serious. Wholly disarming it, however, requires exploring Merleau-Ponty's metatheoretical commitments, which I discuss in section 5.4. Before getting there, I first explore the objection that phenomenology's androcentrism arises out of its basic reliance upon situated experience to provide normative guidance about the meanings available for expression. Although Merleau-Ponty doesn't *obviously* universalize any such guidance, insofar as he may provide the resources to navigate the Scylla and Charybdis of correlationism and naturalism, I argue, Merleau-Ponty *must* retain recourse to nondiscursive grounds of experience which license true expressions of the world's own meanings. I contend, therefore, that concerns about the androcentric misattribution of these allegedly nondiscursive grounds *do* threaten the praxis I defend, because they locate within it tendencies to normalize the androcentric and anthropocentric heritage that I employ phenomenological tools to scrutinize. The remainder of the chapter is concerned with responding to this narrower correlationist charge.

I contend that, although Merleau-Ponty himself makes numerous errors concerning the "anonymous" body schema which cannot be entirely disarmed in their own right, they are subject to correction from within the wider Merleau-Pontian schema. I argue that we should understand Merleau-Ponty's own theoretical failures as part of a greater contribution to an intersubjective hyperdialectic which aims to discern nondiscursive grounds of experience shared between situated body-subjects. Moreover, I argue, this sort of hyperdialectic may also uncover problematic sediment that would otherwise remain invisible to its proprietor. Understanding the philosopher's metatheoretical role as an interlocutor within this wider hyperdialectic mitigates the charge of androcentrism and further supports the suitability of a radically reflective ecophenomenological praxis to address nonhuman others in their alterity.

5.1 PHENOMENOLOGY AS FOUNDATIONALIST CORRELATIONISM

In chapter 2, I argued that key ecofeminist insights should be reconfigured ecophenomenologically. The grounds of this claim lie in Merleau-Pontian ecophenomenology's unique ability to address more-than-human entities in *their* alterity. I argued that, particularly because of their focus on the contribution of motor intentionality to the worlds we differentially inhabit, Merleau-Pontian ecophenomenologists may engage with more-than-human entities in an appropriately radical fashion that is incompatible with the naturalism found in much ecofeminist literature.

These claims might seem odd given the hostility feminist philosophers have historically displayed toward phenomenology. Phenomenology isn't up to the task of addressing alterity, many claim, because it relies upon the universalization of certain privileged experiences or "essential" structures of consciousness as *the* foundational norms by which "the things themselves" are revealed.[1]

In some cases, their concerns appear well founded. I'm thinking, for example, about Husserl's *transcendental reduction,* under which "I lose nothing of that which existed for me naively, particularly that which manifested itself as really being. On the contrary . . . *I come to know the world itself,* only now I come to know it *as it always was for me* and fundamentally had to be for me: as a transcendental phenomenon" (Husserl, quoted in Urban 2016, 466). In Husserl's view, Petr Urban explains, "transcendental phenomenology is not an introspective study of consciousness, but 'a radical enquiry into the world'" (2016, 466). However, as is widely acknowledged, Husserl often conflates the character of "transcendental" phenomena with that of objects dictated by Gramscian "common sense."

Here's a salient example. Husserl claims that "each intentional experience is either an objectifying act or has its basis in such an act" (2001, 167).[2] He goes so far as to claim that the "intentional and epistemic essence[s]" of acts by which objects are revealed must be transhistorical and stable, and must hold for any "minded" creature which "understands" the world it inhabits (2001, 314–16). Husserl thereby subsumes operative intentional norms under specific norms of act intentionality. Act intentionality is concerned with thematizing judgments about how things appear to consciousness. However, as Merleau-Ponty's beetles demonstrate, operative intentionality may lack an intentional *object* over and above the intentional act itself.

By making objectification a transcendental condition of experience, Husserl gives act intentionality a certain intellectualistic priority over what appears to be the ordinary mode of engagement for dung beetles or earthworms, and which also underpins human perception. Husserl also thereby reveals a lack of critical self-reflexivity about the intentional salience of his own situated embodiment, smuggling in important value-judgments about what is "essential" or "primordially" available to experience and, correspondingly, to "the things themselves." Despite his protestations, Husserl's account is arguably infected, for instance, by sediment of the mind/nature dualism which unduly grants the faculty of "understanding" only to conscious "subjects" *defined* in opposition to "objects." Under this schema, (neo-Cartesian) consciousness is distanced from material embodiment, rendering Husserl hostile to recognizing heterogeneous nonhuman intentionalities from the outset.

Moreover, I've argued that the disembodied conception of mind seemingly employed in Husserl's *transcendental* reduction is both androcentric and problematically aligned with the project of anthropocentric mastery that ecophenomenology aims to disrupt. Husserl's latent dualism is exemplified by the fact that he takes *objectification*—the mode of revealing perhaps most detrimental to positive engagement with more-than-human entities—to be *the* primordial one by which essences are revealed. If so, then Husserl's oversight is a major one. As Val Plumwood argues, however, such oversights are easier to gloss over for those occupying master perspectives, since their political privilege makes them less vulnerable to the negative effects of the violence that objectification licenses. "Non-Western" subsistence farmers (particularly women) and nonhuman animals are, after all, hit harder by anthropogenic desertification linked to "excessive resource consumption" than are those invested in "Western" agribusiness.

Husserlian errors like the above, the argument goes, are symptomatic of a broader problem with phenomenology: phenomenology takes "how the world *always* was for me" (or the reflective idealization thereof) to be the neutral ground of epistemology and ontology. Phenomenology, therefore, takes insufficiently seriously that any reduction or reflection is always concretely situated, rendering it prone to neglect the sedimented assumptions and commitments partly constitutive of its revelations. Phenomenology allegedly thereby retains the same kind of dualistic heritage that proves hostile to those marked as "other" by the very structures we aim to address, because those Others deviate from the norms (as they apparently *always* were) of the occupied center. Phenomenology, however, may be worse situated than its nonphenomenological rivals to attend to errors of onto-epistemological hegemony. This is because phenomenology seems to be "correlationist" in the sense that there is nothing concrete outside one's grounding phenomenology (such as a naturalistic thing-in-itself) that might provide a yardstick of truth with which to hold one's experiences to account.

Phenomenology's worries may not end there. We've seen phenomenologists like Merleau-Ponty concede that one cannot make another's consciousness entirely present to oneself without collapsing the very alterity to which one intends to do justice. As Husserl's case shows, in attempting this feat, one would succumb to Levinas's objection that phenomenology requires the "reduction of the other to the same." The upshot of Merleau-Ponty's concession, however, is that one cannot simply empathetically "check" one's experience for fidelity with Others' in one's attempt to discern the meanings licensed by, say, a "depleted" rainforest.

Nor can one just *ask* Others to determine which meanings are phenomenologically basic. Jill W. Scott contends, for instance, that, in even the human case, one cannot unproblematically "appeal to experience as uncontestable evidence and as an originary point of explanation—as a foundation upon which analysis is based" (1992, 24). Doing so, Scott argues, makes visible only *situated* experience, actively *concealing* the sociopolitical sediment that betrays "the constructed nature of experience" (1992, 25). Thus, the testimony of Others—even those outside the master perspective—remains prone to mistaking hegemonic sediment for the world's essential meanings. We saw in chapter 2, for example, that some early ecofeminists mistook situated neo-Cartesian assumptions for essential features of their subjectivity and experiences of being "closer to nature" than men. Likewise, Luce Irigaray contends that liberal feminism's perennial problem is that it describes "what a woman is within the horizon of a male subject's culture" (1996, 61).

Jennifer Saul's (2013) research on implicit bias supports Irigaray's contention. Women, it suggests, are equally likely to perceive women's academic work to be inferior, even when the only difference is the name of the putative author. Body-subjects from ethnic minorities also appear no less quick to match negative adjectives to black faces. If, as I've suggested, these biases are underwritten by dualistic assumptions which also license denigration of the more-than-human world, then these too will be similarly hidden from the ecophenomenologist.

Understood as I put it above, by purporting to grant the essential foundations of knowledge, phenomenology ends up being not only anthropocentric, but androcentric, Eurocentric, heteronormative, and so on.[3] Furthermore, because phenomenology is fundamentally introverted in deferring to lived experience for criteria of truth, it isn't clear that tools can be established within phenomenology to dig itself out of the solipsistic trap in which it is apparently mired. If so, then my optimism about ecophenomenology's potential to disrupt our submerged mass would be ultimately misplaced. Phenomenology, the argument goes, only really grasps other entities according to an androcentric purview which phenomenologists are wont to impose upon nonhuman entities in a suspiciously (ontologically) anthropocentric, top-down manner. This is exactly what appears to happen in Husserl's case.

It's important to emphasize that, unlike those addressed in the preceding chapter, our present objection doesn't diagnose any in-principle incompatibility between (eco)phenomenology and (more-than-human) alterity. I've argued that the fact that phenomenology begins from a situated perspective needn't *entail* solipsism or idealism unless one already adopts a set of radical (e.g., subject/object, self/other, or mind/nature) dualisms. This narrower charge holds that phenomenologists are, in practice, hostile to (more-than-human) alterity because phenomenology is problematically predisposed to -centric essentialism and is poorly equipped to address that failing.

5.2 A RADICALLY ANTIFOUNDATIONALIST MERLEAU-PONTIAN RESPONSE

Nevertheless, Merleau-Pontian ecophenomenology isn't *obviously* vulnerable to such onto-epistemological impositionism because, beyond its contention that perception is irreducibly embodied, it makes no explicit claims about stable, transhistorical, or desituated (structures of) experience. Merleau-Ponty continually problematizes "high-altitude thought" and the related objectivistic project of transcendental abstraction. Indeed, I've argued that Merleau-Pontian ecophenomenology's value lies largely in its appreciation of situated bodies and their differential contributions to the world's divergent meanings. Thus, Merleau-Ponty would resist much of the Husserlian dogma above, especially the point about the universal "essences" of phenomena as presented to (what are genuinely heterogeneous and ambiguously) "minded" entities.

Moreover, radical critical self-reflexivity about habitual sediment is focal for Merleau-Ponty. Critical self-reflexivity is also required about the revelations yielded by radical reflection itself (and any reflective distortions therein), which is why philosophy must be *unrestingly* hyperdialectical. This requirement is made clear in the preface to the *Phenomenology*, where Merleau-Ponty contends that *in* reflection, one cannot escape the world *of* reflection, so "Husserl's essences are destined to bring back all the living relationships of experience, as the fisherman's net draws up from the depths of the ocean quivering fish and seaweed" (*PP*, xv). It seems inevitable, then, that the sediment of one's multifaceted situation will imbue the background of any account of self-knowledge or purportedly "primordial" revelations about the more-than-human world. As Merleau-Ponty famously writes: "Man is a historical idea and not a natural species. . . . There is in human existence no unconditioned possession" (*PP*, 170).

By understanding embodied existence within its "concrete historical situation," rather than through a wholly transcendental lens, or by adopting the mechanistic framework of "naturalistic ideology," Judith Butler notes, Merleau-Ponty *appears* to provide a nonreductive, radically antifoundationalist, existential phenomenology (1989, 85). But, as Butler recognizes, this interpretation is, in the final analysis, implausible. While Merleau-Pontian phenomenology may be nonreductive, it cannot be radically antifoundationalist. Contra Butler, however, I will argue in what follows that Merleau-Ponty must reject radical antifoundationalism *because* of the seriousness of his engagement with the fundamental embodiment of intentionality.[4]

5.2.1 Oksala's Model

To clarify some problems with the radically antifoundationalist interpretation, it will be instructive to address a specific formulation of it. Johanna Oksala (2006) advances one such model. Oksala begins by conceding that the body schema cannot be inessential to the Merleau-Pontian picture. We've seen that one's body schema is intimately bound to one's expressions of the world because one's bodily abilities and dispositions delineate its intentional physiognomy. The body schema is also essential to the potential for non-anthropocentrism within Merleau-Pontian ecophenomenology. By emphasizing the contribution of one's bodily abilities and embodied expectations to meanings perceived, the body schema disrupts the purported neutrality of the dualistic and objectivistic revelations licensed by scientific naturalism. Thus, the body schema brings into clear focus the fact that our worlds are tied to the bodies that we are, and that our bodies, and therefore our worlds, may differ in dramatic respects.

However, like Jean Grimshaw, Oksala denies that, even in the human case, Merleau-Ponty *need* attribute "some kind of ontological priority" (Grimshaw 1999, 103) to any particular body schema—and, thus, one mode of revealing—over another. The "foundationalist" interpretation that Oksala rejects is "a philosophical position that seeks in an analysis of the body's structures a universal and stable foundation for subjectivity" (2006, 211).[5] Oksala makes observations similar to Sara Heinämaa's about the ubiquity of intentional heritage to argue that, for the suitably critical Merleau-Pontian, there is no essential or presedimentary reality *at all*. In Oksala's reading, "transcendental intersubjectivity—language, tradition, and community—is understood as *the reality-constituting principle* providing the conditions of possibility for all forms of subjectivity as well as objective reality" (2006, 218; emphasis mine).

Oksala also argues that her interpretation needn't entail solipsism or disengagement with Others in the manner that proves problematic for absolute models of alterity. She focuses upon Merleau-Ponty's claim that sharing a world with Others entails its irreducibility to one's *own* "intentional threads" because "the philosopher cannot fail to draw others with him into his reflective retreat" (*PP*, 361). In Oksala's reading, something's meanings cannot be reduced to one's own if experienced by another consciousness within a common *cultural* framework. Thus, Oksala argues, situated intersubjectivity is the basis of "objective" reality, and "social normality" (2006, 220) dictates our intersubjective accounts of it. If "social normality" is, therefore, the primary bearer of meaning, Oksala argues, Merleau-Ponty's account becomes reconcilable with Butler's insofar as body-subjects aren't "historically situated" within a world; both body-subject and world are *historically constituted* (Oksala 2006, 221).

5.2.2 Nondiscursive Grounds of Experience

I will set aside several exegetical problems with Oksala's interpretation since our present enquiry concerns the plausibility of a Merleau-Pont*ian* account rather than the faithfulness of any given interpretation. Oksala's biggest problems lie in allying experience too closely with discourse. Here, I take my cue from Linda Martín Alcoff, who argues that poststructuralist feminism's major error is to think that "experience and subjectivity are produced through the interplay of discourses" and, therefore, "the ultimate source of knowledge about social meanings cannot be 'experience,' or, worse, 'perception,' but language and textuality" (2000, 252).[6] Under this reading—to which Oksala explicitly subscribes—phenomenology is robbed of its epistemic primacy. Lived experience becomes discursively constructed rather than an irreducibly basic means of revealing the world's meanings.

However, as Alcoff argues, poststructuralists (like Butler and Scott) here invoke a false dilemma between an "epistemology of experience" and an "epistemology of theory" (2000, 254). This dilemma is invoked because of simplistic analyses of the ilk Oksala's (2006, 221) exemplifies, where *either* "conditions of possibility for perception" and subjectivity are structured in the same "ahistorical or universal" way that Husserl implies and Scott decries (and one adopts the former, naive, epistemology), *or* perceptual schemas are "dynamic and developing structures derived from our cultural environment" (and one adopts the latter epistemology, where the [curiously disembodied and desituated] sort of "theory" Oksala advances alone explains experience and subjectivity).[7] However, as Alcoff argues, "one need only have recourse to Hegel's concept of *Erfahren* to develop an alternative account that understands experience as epistemically indispensable, but never epistemically self-sufficient" (2000, 254). Given his account of the dynamic ambiguity of the body schema, Merleau-Ponty would surely concur: "It is impossible to superimpose on man a lower level of behaviour which one chooses to call 'natural,' followed by a manufactured cultural world. Everything is both manufactured and natural in a man, as it were, in the sense that there is not a word, not a form of behaviour which does not owe something to purely biological being—and which at the same time does not elude the simplicity of animal life" (*PP*, 189).

If our two epistemologies were mutually exclusive, then attention to surreptitious historico-cultural sediment would motivate adopting a discursive constructivism (or, like Scott, an outright epistemological skepticism) over a naive, ultimately hegemonic, phenomenological account. Alcoff, however, uses the example of sexual violence to demonstrate that it would be a grave "metaphysical error" (2000, 256) to believe that reality is discursively *constituted* in the manner that poststructuralists suggest, given the dilemma's falsity.

Alcoff supports her contention with reference to Michel Foucault's example of a "farm hand . . . somewhat simple-minded," who, in 1867, was detained for life following a sexual encounter in which he offered "a few pennies to the little girls for favours the older ones refused him" (Foucault 1998, 31–32). Foucault contends that this incident marked a watershed in which "these timeless gestures, these barely furtive pleasures between simple-minded adults and alert children" (1998, 32) were forever transformed through ideological discourse. Once supplemented by phenomenological description, however, the role of discourse appears fundamentally overplayed. Alcoff draws upon more recent testimony of children paid for sexual encounters, the horror of which suggests that, given the vulnerability of children and their bodies, the underpinning phenomenology is such that they *couldn't* exist as shared events of simple pleasure. The "meanings," rather, "inhere partly in the embodied experiences themselves" (2000, 269). Alcoff argues, therefore, that there must exist some *nondiscursive grounds* common to embodied experience which reducing experience to discourse would unduly preclude from our onto-epistemological investigations.

Alcoff uses this argument to emphasize, as does Merleau-Ponty, that although, in one's situated experience, one cannot easily differentiate the discursive and nondiscursive, both *grounds* exist. Foucault may rightly observe that discourse has material effects on our identities and the world's physiognomy. Technological Enframing, for instance, is largely discursive and shapes our accounts of which entities exist and how they are properly characterized. Technology may even cause things like "the labor market" to come into existence. However, these admissions cannot legitimate *reducing* the meanings of the Rhine, food waste deposits, glaciers, ocean acidification, or destabilized ecosystems *to* discourse surrounding them. Neither do they mean that, as the example of sexual violence illustrates, there is nothing primordially nondiscursive that really does ground our concrete experiences of them, at least in part.

The real issue is more intricate than the aforementioned false dilemma lets on. Both grounds show up in experience and it's difficult to reflectively tease them apart. Even our pain experiences, for instance, cannot be *entirely* nondiscursive because they are partly delineated by sediment. Addressing the racially motivated assault of a black colleague, Sonia Kruks writes that "I cannot share my colleague's experience fully, but neither is it entirely closed to me" (2006, 44). Kruks's experience isn't infused in the *same* way by the lived experience of racist politics or objectifying prejudice. Nevertheless, since we must emphasize *both* nondiscursive and discursive aspects of revelatory experience, "there is no conceivable alternative basis or ultimate justification for knowledge other than my experience in the world" (Alcoff 2000, 262). Thus, in doing justice to more-than-human reality, we *must* retain an ecophenomenological model which resists making all onto-epistemological judgments contingent upon discourse alone.

But there is another moral to Alcoff's cautionary tale. We've seen that, for Merleau-Ponty, the modes of bodily attunement specific to our embodied situations cannot be entirely separated from the grounds subsequently expressed as the world's meanings. As our marble example in chapter 1 illustrates, body-world dialectics delineate one's self-experience and one's experience of the world. To participate in this dialectic, one must begin with a body of some form or another as one's basic means of being a body-subject in the first place. However, as Jakob von Uexküll's work neatly shows, the body cannot begin as a mere intentional nullity. Bearing this in mind, we can see Oksala's central problem: Oksala fails to recognize that body-subjects must share certain irreducibly nondiscursive commonalities in bodily norms that *allow* nondiscursive grounds of experience to express themselves as they do. There must, in short, be some minimal limits to the discursive plasticity of the body schema that the radical antifoundationalist cannot concede. After all, there does seem to be something about the fundamental contributions of *both* body-subject *and* world that delimits the phenomena experienced by the children to whom Alcoff refers.[8] However, their experiences cannot be wholly "subjective" in any accurate sense because the world itself licenses these (harrowing, painful) expressions through those nondiscursive grounds.

If I'm right about this, then the perceptual anonymity of which Merleau-Ponty speaks must refer to more than the prereflective character of ordinary experience. As I argued in chapter 3, sedimented habits may be anonymous in this sense. One isn't typically attentive to one's objectifying attitude when using exploitatively produced consumer goods, or when one consumes palm oil products without regard for their wider resonances, for instance. But acquired discursive norms cannot *entirely* exhaust embodied experience without distorting either the meanings truly available for expression or our basic means of expressing them. Thus, although the body schema may be dynamic and subject to sediment via the intentional arc, a plausible account of it must retain some structural commonalities amenable to expressing certain nondiscursive grounds of experience.

This contention comes with two caveats. First, I don't mean to risk the homogenizing suggestion that *all* bodies or members of a species possess the same fundamental schematic norms. It's at least plausible that heterogeneous human bodies with differential ways of being-in-the-world may experience different nondiscursive grounds of experience. Irigaray suggests as much when she says: "Men and women are corporeally different. This biological difference leads to others: in constructing subjectivity, in connecting to the world, in relating" (Irigaray 2000, 96). As previously explained, however, I worry about the extent to which Irigaray thinks these bodily differences cause a binary rift in experience. I suspect that, as Kruks's pain example suggests, at least some nondiscursive grounds are shared between (almost) all humans, at least. My concern, as ever, is to do justice to continuity and difference.

Second, I'm not making the standard foundationalist claim that nondiscursive aspects of experience may be revealed only by immutable bodily structures. It's arguably less clear, for instance, that scenarios equivalent to Alcoff's involving genuinely consenting adults *couldn't* present nondiscursive grounds of "simple pleasure" shared between parties. Furthermore, it's plausible, as Kruks (2006, 42) claims, that predominantly nondiscursive meanings (concerning reproductive justice, for example) *might* differ between pre- and postmenopausal women. More work should be done on these and related questions, but it is beyond the scope of this book. My focus *here* is twofold: to emphasize the importance of nondiscursive grounds of experience, and to explore some plausible implications for how we might understand the meanings body-subjects locate within the more-than-human world.

By readmitting nondiscursive grounds of experience, however, the Merleau-Pontian phenomenologist again becomes vulnerable to the charge of onto-epistemological impositionism whereby the product of discursive sediment is taken to be primordially nondiscursive. Our problem, it might be said, merely shifts from allegedly essential structures of consciousness (in Husserl) to the primordial structures of embodied perception (in Merleau-Ponty). I shall return to this issue in section 5.3, but first we must appreciate why nondiscursive grounds are so important for the Merleau-Pontian *eco*phenomenologist specifically.

5.2.3 Nondiscursive Grounds and Ecophenomenology

We've already explored some of the manifold reasons why attention to nondiscursive grounds of experience is pivotal to our ecophenomenological project:

1. Failing to identify nondiscursive grounds undermines the agency and resistance of the more-than-human world, recognition of which is politically significant insofar as it disrupts our illusions of anthropocentric mastery. By reducing experience to discourse, we unduly deny the (quasi-)agentic import of nondiscursive grounds of experience both "external" (e.g., the "arresting" power of refuse waste) and "internal" (e.g., the cognitive effects of fats consumed) to our bodies.
2. The constructivism that Oksala reverts to in lieu of nondiscursive grounds of experience relies upon a representationalist (or idealist) schema which dualistically cleaves agentic human "culture" (as *discourse*) from nonhuman "nature" (as *resource:* inert and plastic matter). By leaving this dualism intact, we unduly limit the extent to which we may disrupt the colonial attitudes which precipitate our crisis situation.
3. Due to its neo-Cartesian heritage, the aforementioned cleavage directs us toward theoretical positions which underplay their embodied and situated nature. This is problematic because (as various speculative realisms demonstrate) by detaching theory from one's phenomenological opening, one becomes more prone to anthropomorphize the meanings that "reality" yields.
4. This misleading desituation of theory also veils the material relationality in which our embodiment implicates us. As Bennett shows, by failing to address our material relationality, we're more prone to hyperbolize our autonomy and to further misconstrue the extent to which our attitudes and behaviors may impact (somewhat unpredictably) upon the more-than-human world, further perpetuating unsavory illusions about anthropogenic possession and control.
5. The identification of nondiscursive grounds allows the Merleau-Pontian ecophenomenologist to evade correlationism and the problematic environmental consequences licensed by species solipsism and/or de facto idealism. Nondiscursive grounds are our unmediated point of contact with the world itself. By separating the body schema from them, Oksala (2006, 219–20) explicitly reduces the view-from-everywhere to a reality intersubjectively *constituted* by

humans in the manner problematized by Quentin Meillassoux and Graham Harman.

6. Perhaps most important, nondiscursive grounds facilitate radical reflection. Radical reflection, recall, consists of two strands. The first seeks to uncover the meanings that the world licenses in situated expressions via the basic intentional structures one coconstructs with it. I think it's now plausible to describe this strand in terms of attention to *nondiscursive* grounds of experience. Radical reflection's second strand seeks to identify where sediment or reflective distortions gain misleading onto-epistemological status. We can now appreciate that this second strand of radical reflection has two roles: attention to *discursive grounds of experience,* and attention to *nondiscursive grounds of experience which lack universality*: where the intentional grounds one extends to nonhuman entities like beetles, forests, or glaciers are narrowly anthropomorphic, and so on.

Permit me to return to a few examples. When reflecting suitably radically, one realizes that the sky's phenomenal blueness isn't straightforwardly either volitional or acquired; its meaning is nondiscursively driven by the world itself. The sky is unavoidably blue, for perceivers with visual systems that can see it; hence its blueness is given anonymously in human (but not canine) experience. Given that radical reflection reveals extensional realism's metaphysical stipulations to be more significantly discursive, however, we shouldn't see that lens as similarly fundamental. Given the parameters of certain regional "cuts" (e.g., given certain normative assumptions about the relevant observational data and exclusionary measurement apparatuses), it *may* really be true—as John Baird Callicott claims—that ecosystems are reducible to energy flows. However, as I argued in chapter 3, mistaking this homogenizing characterization for *the* reality of ecosystems gives undue priority to discourse via objectivistic or dualistic sediment.

If we're to do justice to the irreducibly revelatory value of experience which Alcoff foregrounds, then some sort of onto-epistemological priority must be given to its nondiscursive grounds, especially where one hopes to do justice to *more-than*-human alterity. Moreover, if we're to problematize hegemony through our guiding view-from-everywhere, we must be attentive to different sorts of grounds as they contribute to experiences. Failure to carry out this reflective task sufficiently radically was both the naturalists' and the speculative realists' major error.

I should nevertheless emphasize that I don't mean to imply a value dualism between discursive (bad) and nondiscursive (good). I've already argued that both operate only as *grounds* of experience and one is never entirely free from the other at the level of full experiential expressions. Moreover, it's evidently *essential* to a well-informed view-from-everywhere that ecophenomenological analyses engage dialectically with natural and social sciences which operate at various levels of discursive abstraction from the phenomenological openings which are their soil. Furthermore, in some cases, it's the reflective lens of science that allows it to discern latent discursive sediment in our phenomenologies. I'm thinking here about the research in nonequilibrium thermodynamics which informs Bennett's decentralization of agency. Similarly important breakthroughs are being made in "4E" cognitive science, not least because 4E approaches don't adhere to objectivistic or representationalist schemas quite as stringently as classical cognitive science.[9] Dialogue with these disciplines can, and must, inform a radically reflective ecophenomenological praxis.

Our issue isn't the axiological status of (non)discursive grounds as such, but how (non)discursive grounds contribute to our concretely situated phenomenologies and any abstractions subsequently licensed. We need to be able to recognize sediment to some degree *as sediment* if we're to be able to disrupt the dominance of a praxis of objectification, because mistaking sediment for a primordial mode of revelation is what ultimately licenses the colonial errors we discern in Bacon, Descartes, Crutzen, Husserl, and Harman.

Where does this leave us? We've seen that, while Merleau-Pontian ecophenomenologists needn't be guilty of Husserl's essentialism, they must nevertheless affirm the existence of nondiscursive grounds of experience. Our question now becomes the following: Are Merleau-Pontians prone to conflate the respective grounds of experience in an imperialistic manner that they aren't sufficiently well equipped to address? If so, then the objections surveyed above might prove fatal for an ecophenomenological praxis that seeks to disrupt the colonial violence of our crisis situation through a radically reflective habitus of critique.

5.3 (NON)DISCURSIVE GROUNDS AND THE BODY SCHEMA

Let's begin with some objections to Merleau-Ponty's core account of the body schema. Iris Marion Young and Frantz Fanon have each sought to show that Merleau-Ponty's basic account of the "anonymous" (i.e., fundamentally nondiscursive) body schema betrays the sediment of Merleau-Ponty's own privileged (white, male, "Western") situation. Their charges, if successful, speak against Merleau-Pontian phenomenology's ability to engage with radical alterity, because Merleau-Ponty would be shown to be prone to mistake the roles that (non)discursive grounds play in even our most basic self-experience. Moreover, if successful, they suggest that Merleau-Pontian phenomenology's inveterate androcentrism—let alone its value-laden ontological anthropocentrism—situates it poorly to attend to this error. If Fanon's and Young's objections undermine Merleau-Ponty's openness to the alterity of relatively similarly embodied beings (i.e., other humans), they also problematize Merleau-Pontian *eco*phenomenology's capacity to institute an anticolonial view-from-everywhere in which radically heterogeneous entities are involved on their own terms. I shall begin with Young.

5.3.1 The Masculine Body

While Oksala is right that Merleau-Ponty doesn't ascribe foundational priority to any particular body schema, Young's objection targets Merleau-Ponty's contention that, regardless of its specific character, the body schema is *unified* in all normal cases. For Merleau-Ponty, the synthesis of one's bodily abilities through the body schema empowers one to appropriate one's world. One remains "in undivided possession" of one's body "through a body schema in which all [of its elements] are included" (*PP*, 98; translation altered). Merleau-Ponty's "anonymous" body schema is thereby characterized as something nondiscursively lived through as "pure fluid action" (Young 1989, 59). The "ambiguous transcendence" typical of "feminine" motility, however, problematizes this contention (Young 1989, 54, 58).

Young draws upon empirical research on "feminine" gaits and throwing styles to argue that women often live their bodies in a more fragmented sense than Merleau-Ponty allows because women experience their bodies largely as burdens and, therefore, more significantly as objects (*Körper*) than men do. This is because, Gail Weiss argues, women differentially "mediate their own relationship with their bodies as they are seen by others" (1999, 46–47). In a misogynistic, sexist, or dualistic society, Sandra Bartky (1990) explains, the panoptical societal gaze is differentially present in women's lives. As a "female" body-subject, one is officially afforded subject status. Yet, largely because of the dualistic sediment which imbues "Western" society, a woman is constantly and forcibly reminded that she is an object in ways that men, because of their privilege, are spared. This might be in terms of her casual sexual objectification, the differential importance applied to her physical appearance or reproductive capacities, and so on. These pervasive acts of objectification begin in childhood and ensure that "the woman herself often actively takes up her body as a mere thing" (Young 1989, 66), causing a rupture within her relationship with her body that promotes the bodily disunity typical of feminine comportment.

This differential objectification also seems to bear directly upon the meanings that the world presents through the modality of "inhibited intentionality" (Young 1989, 59). For Merleau-Ponty, one's well-integrated body schema licenses a fundamental "I can" which "opens [one] out upon the world and places [one] in a situation there" (*PP*, 165). The manifestation of one's bodily capacities inherent to this "I can" operatively delineates the nondiscursive meanings that the world "teaches" us. However, Young argues, by living one's body as a burden, one prereflectively underestimates one's bodily capacities. Worldly affordances are, therefore, often grasped such that, while "*one* can," as a situated *female* body, "I cannot" (1989, 60). The expressive unity that Merleau-Ponty naturalizes between one's bodily capacities and worldly possibilities thus misrepresents the "discontinuous unity" (Young 1989, 61) of "feminine" intentionality. Since, for the Merleau-Pontian, one's worldly grasp is intimately related to how one comports one's body within it, "feminine" worlds must be configured significantly differently from masculine ones in a manner left unattended by the "anonymous" terms of Merleau-Ponty's body schema.

Young's overall objection is to the transhistorical, *abstract,* character of the body schema which, as Merleau-Ponty mistakenly presents it, affords the nondiscursive grounds of experience we seek. Merleau-Ponty argues that "in order that we may be able to move our body towards an object, the object must first exist for it, our body cannot belong to the realm of the 'in-itself'" (*PP*, 139), but each of Young's observations calls this claim into question. For "feminine existence," Young argues, "the body frequently is both subject and object for itself at the same time and in reference to the same act" (1989, 61). The feminine body is fragmented, and is *lived*, in part, as an object *just like any other.* Thus, Young suggests that the "nondiscursive" grounds of experience licensed by Merleau-Ponty's body schema misrepresent their origins, because they reflect the androcentric phenomenology of the occupied center. Furthermore, through Merleau-Ponty's inability to recognize radically different schemas like these "feminine" ones, he displays a tendency to homogenize allegedly nondiscursive grounds in a manner hostile to even human alterity. Merleau-Ponty is thus dangerously close to repeating Husserl's error of onto-epistemological imposition.

5.3.2 The White Body

Fanon's account of being a black body-subject in a racist society shares striking similarities with Young's. Helpfully, however, Fanon relates his experience to the same master/slave dialectic that Merleau-Ponty invokes. Fanon takes up Hegel's notion that "self-consciousness exists in and for itself when, and by the fact that, it so exists for another; that is, it exists in being acknowledged" (Hegel 1977, 111). The standard Kojèvian interpretation holds that the intersubjective struggle is reciprocal and that even the slave's transcendence is recognized. However, Fanon argues, the black person, dualistically *identified* with certain physical features, is precluded from a *reciprocal* struggle because his appeals for recognition are ignored: "For not only must the black man be black; he must be black in relation to the white man. . . . The black man has no ontological resistance in the eyes of the white man" (2008, 83). Defined by his racialized appearance in the gaze of Others, Fanon cannot substantively exercise his transcendence because he is "locked into his body" (2008, 175). This explains how Fanon becomes an object in much the same way as the "feminine" body-subject.

Jeremy Weate notes that, since the above dialectic prevents his "free agency" and construction of a "self-image," the black person cannot construct his own body schema; a "white *mythos*" instead "inserts itself between the black body and its self-image" (2001, 7). Fanon thus describes sketching *below* his body schema a "historico-racial schema" (2008, 84): the sedimented product of history as constructed by white people, in which he is trapped. Fanon describes "discovering [his] blackness" as being "battered down by tom-toms, cannibalism, intellectual deficiency, fetichism [*sic*], racial defects, slave-ships, and above all else, above all: 'Sho' good eatin' '" (2008, 84–85). Fanon's body schema finally "crumbles" and is replaced by a "racial epidermal schema," in which the historico-racial schema "becomes 'naturalised' as a *condition* of skin" (Fanon 2008, 84; Weate 2001, 10). Unable to freely encounter the world with a primordial "I can," this schema in turn ensures that the black person in a racist society is laden by a "nauseating" responsibility for "his" "race" and "ancestors" (Fanon 2008, 84).

Fanon poses a problem for Merleau-Ponty because, in Merleau-Ponty's master-slave dialectic, "the other's gaze transforms me into an object, and mine him, only if both of us withdraw into the core of our thinking nature, if we both make ourselves into an inhuman gaze" (*PP*, 361). As a matter of inescapable perceptual faith, Merleau-Ponty claims, one must experience domination as a fundamentally *embodied* relation. In episodes of violence—even the violence of the gaze—other people (nondiscursively) present as suffering subjects in whom—*via our common embodiment*—we're partially implicated. Through the gaze, therefore, "the other is for me no longer a mere bit of behaviour in my transcendental field, nor I in his; we are collaborators for each other in consummate reciprocity" (*PP*, 354). Like the "feminine" body-subject, however, the black person's body apparently *prevents* him from resisting his objectification. Thus, the free and reciprocal body-world dialogue that Merleau-Ponty thinks is spontaneously available to all body-subjects seems instead to be contingent upon the modes of embodiment licensed by one's sociohistorical situation. That is to say, the requisite transcendence to construct Merleau-Ponty's "basic" body schema appears dependent upon how the gaze is differentially administered. "Fanon's critique of phenomenology teaches us that the universal is the end of the struggle, not that which precedes it" (Weate 2001, 17).

By misrepresenting the basic structures of embodied experience, Merleau-Ponty inadvertently highlights grounds for concern about phenomenology's ability to attend to "the first philosophical act": to "restore to things their concrete physiognomy, to organisms their individual ways of dealing with the world, and to subjectivity its inherence in history" (*PP*, 57). This worry only deepens when we remember that we're attempting to address the radical alterity of nonhuman entities, some of which cannot straightforwardly "gaze" at all, and so appear to be at further risk of dualistic objectification. Merleau-Pontian ecophenomenology's anticolonial credentials may be in jeopardy.

5.3.3 An Obvious Response

It's tempting to think that Merleau-Ponty's error is less grave than it appears. While Fanon and Young may expose Merleau-Ponty's limited capacity to identify experiential commonalities (the response might go), this admission little weakens the Merleau-Pontian phenomenologist's ability to discern nondiscursive grounds of experience *as such*. Why? Because they each appear to be describing what Young calls a "handicapped" (1989, 65) schema. Young's "feminine" body schema appears to have been damaged by the patriarchal gaze. Likewise, Fanon claims that "as long as the black man is among his own, he will have no occasion, except in minor internal conflicts, to experience his being through others" (2008, 82). Had they not been contingently subjected to dualistic objectification, they may each have retained the same basic, unified, body schema Merleau-Ponty describes.

We should have little time for this response, however, primarily because it is both *false* and *dangerous* to normalize the body schema in this manner. On the first count, Richard Shusterman accuses Merleau-Ponty of effectively advancing a "misleading binary contrast [between] 'normal' people whose somatic sense and functioning is totally smooth, spontaneous, and unproblematic [and] the abnormally incapacitated: patients such as Schneider who exhibit pathological dysfunction and are usually suffering from serious neurological injury (such as brain lesions) or grave psychological trauma" (2005, 166). Most of us live *between* these binaries. However, "Merleau-Ponty implies that if we are not pathologically impaired like Schneider and other neurologically diseased individuals, then our unreflective body sense is fully accurate and miraculously functional" (Shusterman 2005, 166). Shusterman uses trivial examples, such as the commonplace mistake about the position of one's head which prevents an effective golf swing, to show this to be straightforwardly false.

Likewise, Grimshaw (1999, 107) highlights the inhibited movements of male colleagues in aerobics classes to argue that Merleau-Ponty's idealized body schema simply isn't realized, even by affluent white men in patriarchal societies. Where these body-subjects *do* get close to spontaneously maximizing their capacities in the requisite sense, Grimshaw observes, is invariably in arenas discursively coded "male," like team sports. If the relative inhibition of transcendence that body-subjects experience is intimately related to their differential immersion in such arenas (e.g., women in aerobics, or men in team sports), then such schematic norms appear to be acquired, not primordial.

More problematic for the *eco*phenomenologist, however, is Grimshaw's assessment of the *dangerous normative role* played by Merleau-Ponty's stipulations. Grimshaw notes that Merleau-Ponty's "normal" body schema is principally defined by its lack of inhibition. Inhibition in the body schema, according to the proponent of the "obvious response," results from dualistic discourse that draws undue attention to one's object-body. Grimshaw (1999, 107) argues, however, that this characteristic lack of inhibition mirrors the "masculine" impetus to occupy maximal space which often inhibits Others in ways that (because of *its* privileged status within dualistic discourse) we aren't well positioned to recognize. Grimshaw qualifies her claim with reference to parallels between the "masculine" obsession with noisy and powerful vehicles and an intrusive "masculine" gait or choice of leg position when seated. But it's exactly this expansionist drive that we identified, in previous chapters, as an important aspect of our unreflective commodification and appropriation of the more-than-human world. This sort of drive appears complicit in the reckless consumerism that Bennett seeks to challenge through attention to nonhuman (quasi) agency, for instance. The uninhibited intentionality that Grimshaw problematizes is also particularly poorly positioned to *hesitate* in the requisite manner to disrupt the uncritically objectifying praxis which Erazim Kohák equates with "urban experience."

If, as Shusterman, Grimshaw, Young, and Fanon each imply, the "nondiscursive" boundaries of Merleau-Ponty's body schema reflect a misleading ideal of bodily communion with the world which betrays the sediment of a constructed "master" position, then Merleau-Ponty's nonanthropocentric and nonandrocentric credentials are under threat. Not only may the allegedly nondiscursive grounds of experience manifested through Merleau-Ponty's body schema be more significantly discursive than they appear, some of the discursive sediment thought fundamental to it appears problematically aligned with a propensity toward colonial violence. In support of this claim, we saw in chapter 3 how Merleau-Ponty himself links lack of inhibition to colonial habits. For Merleau-Ponty, recall, this is because when one's tools work well—when one's appropriation of worldly things proceeds largely unimpeded, as when men are not primed to focus on their "objective" bodily limitations—there exist insufficient moments of friction or denial (what Alia Al-Saji calls "inassimilable events") to promote critical self-reflexivity to sediment or the ways we appropriate the world to *our* goals. This thought was behind our recommendation of the virtues of attention and hesitation (and Kohák's less convincing imperative to "take to the woods"). By conceiving the body schema as he does, Merleau-Ponty appears to actively *limit* critical self-reflexivity as a matter of praxis.

Let's recap. We now have good reason to think that Merleau-Ponty misidentifies the relative grounds of experience licensed by his "basic" body schema. Moreover, the grounds he misidentifies as fundamentally nondiscursive seem to result largely from dualistic discourse and are, therefore, detrimental to uncovering its sediment. Since the promise of Merleau-Pontian ecophenomenology lies partly in its ability to expose such colonial sediment and, subsequently, to contribute to a more positive view-from-everywhere that resists anthropocentric mastery, these admissions are a serious blow. Furthermore, since (as Alcoff argues) nondiscursive grounds may show themselves only in experience, yet (as Young, Fanon, and Grimshaw argue) Merleau-Ponty is prone to mistake dualistic sediment for nondiscursive grounds, our admissions may be problematic for a *praxis* of suitably radical reflection. After all, such a praxis can only really get off the ground when guided by inassimilable events. If phenomenology's androcentrism makes it prone to mischaracterize discursive sediment as the nondiscursive, however, it's unclear that it may retain sufficient openness to those inassimilable events. Thus, even Merleau-Pontian ecophenomenology may ultimately be prone to result in the sort of de facto correlationism hostile to its goals. Understood as such, it's not clear that an ecophenomenological rethinking of forest ecosystems, deforestation, or anthropogenic climate change would constitute an improvement upon the models natural scientists already offer, regardless of the broader range of meaningful phenomena that are amenable to a phenomenological ontology.

5.4 THE DIALECTICAL INTERPRETATION

These are important objections, but the Merleau-Pontian project isn't yet doomed, because the preceding analyses neglect some of the important tools Merleau-Ponty bequeaths us. To utilize these tools, however, our first move must be to concede that Merleau-Ponty distorts the body schema's "nondiscursive character" by presenting it as he does. But this move needn't make Merleau-Pont*ian* ecophenomenology false or unsuitable for our purposes. Why? First, because Merleau-Ponty's own androcentric characterization of the body schema is inessential to his broader (and more plausible) account of embodied perception. The important ecophenomenological insights I've drawn from Merleau-Ponty pivot upon the principle that one's differential embodiment contributes in significant, if often clandestine, ways to the meanings the world affords one's situated "grasp." Even in jettisoning Merleau-Ponty's problematically "masculine" schema this principle remains untouched; it's actually structurally essential to Young and Fanon's objections. Abandoning some of the specifics of Merleau-Ponty's own account needn't motivate a problematic radical antifoundationalism either. As previously noted, there is no in-principle reason why *other* accounts of the body schema (or body schema*s*) couldn't do justice to common nondiscursive grounds of experience without retaining problematically colonial or expansionist drives at their core.

Our next step is to address what, at a metatheoretical level, is going on when Merleau-Ponty proposes his mistaken account of the body schema. To do this, we must address some tensions. The present objection holds that phenomenology involves a sort of de facto correlationism whereby one cannot transcend the discursive sediment of one's experience to arbitrate which grounds go beyond it. This is a contention, however, to which Merleau-Ponty wouldn't be *entirely* adverse. He disavows "high-altitude thought" and thinks that, since experience *always* reflects the sediment of one's situation, a full transcendental reduction is incompletable. But if so, then it appears methodologically *impossible* for Merleau-Ponty to satisfy the task our critic ascribes to him; it would require a project of transcendence of exactly the ilk Merleau-Ponty cautions against. This *cannot* be his project, and Merleau-Ponty tells us as much: "A philosophy becomes transcendental, or radical, not by taking its place in absolute consciousness, without mentioning the ways by which this is reached, but by considering itself a problem; not by postulating a knowledge rendered totally explicit, but by recognising as the fundamental philosophic problem this *presumption* on reason's part" (*PP*, 63).

Predicated upon the same misreading of Merleau-Ponty's project is Shannon Sullivan's contention that, for Merleau-Ponty, embodiment provides a "shortcut" (1997, 8) to Others. In Sullivan's reading, embodiment somehow facilitates privileged access to common "anonymous" structures of experience—transcendentally stripped of the particularities of one's situation—which one may reflectively uncover. However, by beginning from the assumption of harmony or unity between parties, Sullivan claims, Merleau-Pontian phenomenology is inherently hostile to alterity because any intersubjective "dialogue" effectively becomes "a solipsistic subject's monologue that includes an elimination of others in its very 'communication' with them" (1997, 1). Ultimately, Sullivan thinks, the Merleau-Pontian phenomenologist is apt to mistake *one's embodied situation* for *the situation of being embodied*. For Sullivan, therefore, embodiment affords the same "unproblematic" shortcut to the Other that transcendental consciousness does in Husserl, and hence Merleau-Ponty repeats Husserl's colonial error.

However, as Kruks (2006, 37) clarifies, Merleau-Ponty *never* claims that one's body provides a shortcut to the other party. Embodiment is, for Merleau-Ponty, only the "medium of our communication" (*PP*, 181): the source of the *possibility* of a common world in which a dialectic might take place. And, lest we forget, philosophy must be *hyper*dialectical because

"Hegel is the museum." We saw in chapter 3, for instance, how *sustaining* a radically reflective hyperdialectic with the more-than-human world is essential to a praxis which resists appropriating it under problematically objectivistic terms. This is because one must remain affectively open to the surprising meanings the world throws up (e.g., that trees communicate, that rubbish may be somehow agentic, that objective thought isn't ontologically basic, or that science's positivistic terms needn't exhaust reality), attending to them in turn. For Merleau-Ponty, "the core of philosophy," therefore, "lies in the perpetual beginning of reflection" (*PP,* 62).

Of course, for Merleau-Ponty, one's embodiment provides the necessary ontological continuity for any genuine dialectic to get started. Without it, one really might be trapped within (species) solipsistic perspectives *upon* the world. However, dialectical relationships are always both sustaining *and* denying. Overlooking this point is the source of Sullivan's error. "The dialectic," Merleau-Ponty explains, "is not a relationship between contradictory and inseparable thoughts; it is the tending of an existence towards another existence which denies it, and yet, without which it is not sustained" (*PP,* 167–68). If dialectics between body-subject and world were merely sustaining, radical reflection needn't be a perpetual beginning; the world would simply present its meanings to suitably engaged individuals. As we've seen Merleau-Ponty argue, however, this kind of thinking reflects a lack of attention to one's situated and embodied historicity and to the inexhaustible alterity the world retains. It affords only the "closed" dialectic that typifies scientific naturalism which we disavowed with Donna J. Haraway in chapter 2.

Merleau-Ponty recognizes that the world may dialectically deny us in several productive ways that, if we're disposed to address it *in its alterity,* we must be open to. We've already seen some of them. It may frustrate our embodied expectations (think about the shipwreck or Bennett's rubbish). It may shatter our present grasp through inassimilable events (think about Trewavas and Baluška's plant neurobiology). The world may even withdraw from our grasp and retain a radical excess (think about Uexküll's ticks). The moral here being that, as only a situated "style" of embodiment among others, one cannot freely survey or exhaust the more-than-human world's meanings without collapsing the very alterity to which, in my reading, radical reflection intends to do justice.

Since, for Merleau-Ponty, "there is no rupture between our being in the natural world and our being in the social world, we might expect similar tensions also to inform our relations with other people" (Kruks 2006, 35). In fact, Kruks argues, Merleau-Ponty's entire political *oeuvre* (and especially his experiment with Marxism) is characterized by commitment to an intersubjective dialectic governed by affirmation *and* denial relations. Merleau-Ponty's guiding political question is "*whether* communication might be possible in a world of endemic conflict" and, more specifically, whether communication might *begin* through "embodied intersubjectivity" (Kruks 2006, 39). However, as Fanon's account attests, embodiment is no shortcut to harmony. Kruks thus reads the thrust of Merleau-Ponty's political theory as follows: that *if* we seek to establish positive dialectical relationships with Others, then these must be built upon the grounds that our embodiment affords to access a common world.[10]

This, I think, is Kruks's key insight: embodiment provides only the possibility of a ground upon which *we may build* better dialectics. But since other parties are differently situated, one cannot entirely accurately or reliably prejudge what the other party will affirm, and what they will deny, in advance of participating in the dialectic. In intersubjective experiences, one might have *some* affective awareness of how the Other's pain might feel, or the quasi-familial relationship she might share with a specific area of land, for example. However, given one's Gestalt-situatedness, one cannot know how the world (or even a specific pain) *is* for the Other. Believing that one could was the source of Husserl's major error. Furthermore, given the diacritical intertwining of discursive and nondiscursive grounds in our experiences, nor can one know that Others take up the same putatively nondiscursive grounds at all in advance of the dialectic. After all, since one's being is always being-in-the-world, one cannot entirely transcend one's phenomenological experience. Nor can one, therefore, unproblematically desituate or universalize any theoretical pronouncements derived from it.

This tension between being concretely situated in a sociohistorical life-world and the demand for radical reflection to bring our intentional commitments into view is what Merleau-Ponty calls "the paradox of immanence and transcendence" (*PrP*, 26). In fact, the problem that this paradox unveils exposes a kernel of truth in Sullivan's objection: should one take one's embodiment to provide a transcendental shortcut to the Other's alterity, one *will* become closed off to the possibility of denial essential to a rigorous dialectic. What one needs to do instead is participate in the intersubjective hyperdialectic in earnest. This, I suspect, is what Merleau-Ponty was doing when he made his *provisional* claims about "anonymous" experience.

Taking these insights into account, the interpretation I propose holds that descriptions of the "anonymous" body schema in the *Phenomenology* are contributions to a *hyperdialectical* project, rather than a standardly foundationalist or radically antifoundationalist one. Merleau-Ponty is reflecting upon his own somewhat ambiguous experience in pursuit of shared knowledge about the body as it presents its structure ("anonymously") in perception, rather than either articulating *the* universal account upon which perception is built, or one in which all perception is culturally constituted. The dialectical method promotes open, critical, intersubjective dialogue between radically reflective body-subjects who perpetually propose, reconfigure, and modify descriptive accounts of (prospective nondiscursive grounds of) embodied experience. Intersubjective hyperdialectic at the metatheoretical level simply mirrors the hyperdialectic employed in radical reflection at the theoretical level. The "webs of connections" thereby uncovered afford a common, but unexhausted, *real* world beyond mere construction.

Since transcendental reduction is incompletable, the dialectical method seems to be the only way to establish progressive access to ourselves and a shared world which also takes seriously Alcoff's insight that experience seems to contain something common (or, perhaps, near universal) through its nondiscursive grounds. This interpretation helps us make sense of why Merleau-Ponty seems to describe the foundational structure of every-body, yet, often on the same page, argues that one cannot reflectively transcend one's concrete situation. If Merleau-Ponty is describing aspects of embodied experience that he has reason to think may have common nondiscursive grounds, then the apparent contradictions dissolve. I suspect this was Merleau-Ponty's metatheoretical project all along. Thinking otherwise, to reiterate, would be to neglect the central paradox of immanence and transcendence and consign Merleau-Pontian phenomenology (at least at a metatheoretical level) to Hegel's museum.

This interpretation also explains why Merleau-Ponty's "anonymous" body is sometimes problematically white, "masculine," and so on. Given his own warnings about the implausibility of a complete transcendental reduction, some proposed "commonalities" are bound to reflect historico-cultural sediment more than nondiscursive grounds. Merleau-Ponty *was* wrong about the nondiscursive universality of certain structures of embodied perception. It is, however, perhaps more accurate to say that, metatheoretically speaking, Merleau-Ponty's mistakes reflect his engagement in an intersubjective hyperdialectic in which suggesting, rejecting, and modifying such suggestions in open dialogue with Others is essential to uncovering the nondiscursive grounds of a common experiential world, *without thereby doing undue violence to alterity.*[11]

5.4.1 Beyond Kruks

Kruks's dialectical interpretation helps make better sense of Merleau-Ponty's claims about the body schema, but how does it bear upon our ecophenomenological project? Recall that we've been addressing the charge that Merleau-Pontian ecophenomenology is implicated in a sort of androcentric correlationism. A radically reflective ecophenomenology requires one to identify the respective roles discursive and nondiscursive grounds play in our situated experiences; however, because of its reliance upon experience's revelatory role, ecophenomenology allegedly denies itself the tools to do so. Ecophenomenology is seemingly thus prone to "solipsistically" impose discursive terms upon more-than-human things in the mistaken belief that such terms are fundamentally nondiscursive.

What Kruks's analysis foregrounds is that the charge of solipsistic projectionism is, again, misplaced because it misconceives phenomenology to be a solitary transcendental endeavor. By mischaracterizing Merleau-Pontian ecophenomenology as such, one neglects the possibility that an intersubjective hyperdialectic might provide means of progressively uncovering the (non)discursive grounds of experience we seek.[12] By widening its scope to encompass an understanding of philosophy as *itself* an intersubjective hyperdialectic, Merleau-Ponty affords more realistic opportunities for critical self-reflexivity about our multifaceted experiences and the meaningful phenomena they reveal. As Young, Fanon, Shusterman, and Grimshaw all inadvertently help illustrate, Merleau-Ponty also needn't thereby abandon the radical situation of perspective so important to his characterization of philosophy. By engaging with Merleau-Ponty's body schema and, thus, participating in the dialectical process, each facilitated real progress in uncovering discursive grounds of experience left unidentified in Merleau-Ponty's texts. But this shouldn't be entirely surprising. If, as Heinämaa argues, intersubjective assent relations are an essential facet of one's worldly grasp, our dialectical relationships in "the social world" will *always* play a key role in the establishment of the real.

But Kruks's dialectical interpretation offers more than a relief from solipsism. It affords further *targeted,* intersubjective tools to ensure affective openness to our social and material dependencies that build upon, and go beyond, Al-Saji's recommendation to hesitate. Kruks, therefore, affords further collaborative means of addressing the problematically androcentric, dualistic, and objectivistic commitments and assumptions which undermine more positive encounters with(in) the more-than-human world. To illustrate: note that part of the reason why Fanon's and Young's interjections are so revealing is that they reflect their experiences as members of deprivileged groups. Which isn't to say that, in virtue of certain bodily specificities, they are essentially "closer to nature" or more in-tune with its nondiscursive contours than Merleau-Ponty. Nor, given their complex identities of privilege and oppression, are Young's and Fanon's phenomenological perspectives epistemically innocent.[13] We should heed Scott's warnings about treating them as such. However—and this, I think, is the key lesson of feminist standpoint epistemology—Young's and Fanon's situations better place them to identify and problematize *certain* discursive (i.e., unhelpfully dualistic or androcentric) grounds at play in experience than those more privileged. In Young's and Grimshaw's cases, for instance, the "feminine" body-subject's "inability" to encounter the world with "pure fluid action" instigates an inassimilable event; it draws her attention to the fact that *she* doesn't live according to an uninterrupted modality of the "I can." Young and Grimshaw also thereby allow *us* to realize that this "I can" isn't nondiscursively available to all "normal" body-subjects. Moreover, given its place in a problematically androcentric schema, nor should we wish it to be.

Given the multiplicity of situations and schemas taken up by body-subjects—all of whom share ontological crossover with the rest of the world—we have reason for optimism about the potential for *nonandrocentrism* arising from ecophenomenology's participation in an intersubjective hyperdialectic. To reprise Alcoff's earlier insight, this is primarily because participation in the hyperdialectic affords increased openness to the inassimilable events that disrupt our habitual objectification or androcentric projectivism, helping us, instead, inculcate a radical habitus of critique.

Understanding theory to progress in the social manner Kruks suggests also helps clarify why Merleau-Ponty calls the inherent tension between immanence and transcendence a "fruitful" or "fecund contradiction" (*PrP,* 26).[14] There is some truth to Butler's claim that the attempt to reconcile an abstract subject with a concrete situation as-lived is phenomenology's "impossible project" (1989, 95). For Merleau-Ponty, however, its impossibility doesn't demonstrate phenomenology's futility. The impossibility of any wholly transcendental reflection makes philosophical progress somewhat dependent upon a stance of positive openness to Others and their capacity to deny us. It is, as Jack Reynolds notes, only the ostensible failure of our attempts to overcome the paradox of immanence and transcendence that "wakens us to wonder" (2016, 85).

Thus, through her recognition of this paradox, the Merleau-Pontian ecophenomenologist appears *better* positioned to engage in critical self-reflexivity about her acts of onto-epistemological imposition. Sally Fischer argues, for instance, that, unlike those who adopt the "pathologically detached" stance involved in factory farming or sweatshop labor, Merleau-Ponty's inherently dialectical approach requires "a *general shift* in the way we understand ourselves and live our lives with others, rather than [calling] for an ahistorical set of maxims. It is not just a matter of simply recognizing *that* there is another, with a view different from my own. Rather, ethical recognition is a *mode* of being with another; it *demands* that the other be listened to as *another* possible perspective while still attempting to maintain the goal of genuine reciprocity. . . . There is an obligation, a pact to keep the communicative process *alive*" (Fischer 2007, 210). This shift in *praxis*, Fischer argues, situates Merleau-Pontian ecophenomenology particularly well to take seriously Irigaray's thought that "I am not everything" (1996, 51–52), thereby bolstering its anti-imperialistic credentials. To reiterate: this is because Merleau-Ponty bequeaths to us the tools for a hyperdialectic which is founded upon the nondiscursive grounds afforded by our common carnality, but which, given our embodied situations, *also demands heightened attunement to the possibility of denial*.

Of course, we shouldn't lose sight of the fact that the reason we wish to uncover (non)discursive grounds of experience is, ultimately, to inform a view-from-everywhere encompassing *more-than*-human perspectives. Although far from unrelated, refuting the charges of solipsism and androcentrism won't, therefore, go all the way to securing Merleau-Pontian ecophenomenology's nonanthropocentric credentials in the requisite sense. One remains subject to the paradox of immanence and transcendence, yet one cannot engage in the *same* intersubjective dialectic with the more-than-human world as with other human body-subjects. I say more about this in the next chapter, but by way of preempting my conclusion, I contend that the attempt to do so is actually detrimental to the critically self-reflexive stance that Merleau-Pontian ecophenomenology offers in our quest to disrupt the imposition of onto-epistemological violence.

Even putting this contention aside, however, the fact that intersubjective dialectic doesn't *straightforwardly* extend to all more-than-human entities needn't entail species solipsism. As I've argued, taking alterity seriously means accepting that some of the world's meanings cannot be made available to me. They may simply be denied to me as part of the world's excess. But this isn't the whole story. I've also argued that, as embodied, one is at least partially implicated *within* the perspectives of some more-than-human Others. One may, therefore, make *some* well-justified (if always fallible and provisional) claims about the grounds and meanings available there. What the dialectical interpretation affords, then, are important intersubjective means of promoting critical self-reflexivity about one's violent imposition of discursive grounds (or narrowly anthropocentric, or group solipsistic, nondiscursive grounds) onto other perspectives—*including nonhuman ones*—as if primordial, neutral, or universal. Kruks thus provides nonsolipsistic means of engaging productively and progressively with the more-than-human world that take seriously its capacity to affirm and deny us, and which thereby motivate a stance of openness and humility set against acts of colonial imposition from the outset.

5.5 CONCLUSIONS

In this chapter, we explored the concern that Merleau-Pontian ecophenomenology is poorly equipped to address the more-than-human world on its own terms because it is in some sense both solipsistic and androcentric. More specifically, Merleau-Pontian ecophenomenologists stood accused of problematically mischaracterizing the sediment of "masculine" discourse in ways their philosophies aren't adequately equipped to address. Although I argued that the Merleau-Pontian project *does* require recourse to nondiscursive grounds of experience, this needn't implicate Merleau-Pontian ecophenomenologists in the onto-epistemological imperialism courted by Husserl. I argued that Merleau-Ponty should instead be understood to participate in a metatheoretical hyperdialectic which takes seriously his situated limitations and which, because of its intrinsic, intersubjective focus on openness to dialectical denial, is set definitively *against* the solipsistic imposition of violence with which Merleau-Pontian ecophenomenology is charged. Merleau-Pontian ecophenomenology thus retains its capacity to address more-than-human entities in their alterity and, therefore, retains a pivotal role in addressing the submerged mass of our environmental crisis.

CHAPTER SIX

RADICAL REFLECTION, REVERSIBILITY, AND THE FLESH

"Mister!" he said with a sawdusty sneeze,
"I am the Lorax. I speak for the trees.
I speak for the trees, for the trees have no tongues."
—Dr. Seuss, *The Lorax*

Throughout this book, I've defended the claim that Merleau-Pontian ecophenomenology has a key role to play in addressing the submerged mass of assumptions and commitments which precipitate the environmental crisis. This is primarily because an engaged ecophenomenological praxis may bring our tendencies toward onto-epistemological violence into focus as a matter of habit. I've also argued that Merleau-Pontian ecophenomenology provides useful critical tools to attune us to the meanings that the more-than-human world may license. In the previous chapter, I argued that one such tool comes through the ecophenomenologist's recourse to an intersubjective hyperdialectic. The reason an ecophenomenological praxis requires intersubjective dialogue is that, given one's concrete situation, one will inevitably retain blind spots about one's intentional commitments, even in acts of radical reflection. I argued in the previous chapter that, without the possibility of intersubjective correction—and the corresponding stance of humility encouraged by acknowledging the limits of phenomenological transcendence—the objection that Merleau-Pontian ecophenomenologists are trapped in a sort of de facto androcentric correlationism would carry significant weight.

Nevertheless, the following problem emerges: while an intersubjective dialectic might help diagnose problematically androcentric or andromorphic accounts of embodied *human* experience, *this* dialectic is conducted exclusively by human body-subjects, so its interjections aren't directly subject to correction by the very nonhuman entities whose meanings we seek to accommodate within our view-from-*everywhere*. When I introduced this issue in the previous chapter, I suggested that our worry might be mitigated by emphasizing that, as embodied, the Merleau-Pontian body-subject is partially implicated in the rest of the more-than-human world. It's therefore erroneous to think that humans are bifurcated from the rest of the world in the requisite manner for dialectical engagement with it to become impossible. For Merleau-Ponty, (other) humans are also ontologically connatural with extrahuman entities in the requisite manner to afford (varying degrees of) implication *within* their perspectives. These perspectives may, therefore, also be to some degree intersubjectively clarified.

Even if one accepts my claims about the chiasmic intertwinings between human and nonhuman perspectives, however, one might nonetheless worry that, while one might be implicated in the perspective of a given nonhuman entity with which one is connatural, one might be misled about its meaningful content in a manner that entity cannot *itself* correct. Recall that I previously accused Val Plumwood of anthropomorphically misrepresenting the perspective of a snake on her porch. The snake, however, cannot straightforwardly confirm or disconfirm my suspicions in the same way as Frantz Fanon or Iris Marion Young could. Participation in the intersubjective hyperdialectic simply isn't open to nonhumans in the *same* manner as it is to humans. Thus, one might think that, because of the inherent asymmetry of human-world dialectics, the issues of androcentrism and andromorphism don't parallel the issues of anthropocentrism and anthropomorphism sufficiently closely for an ecophenomenological praxis to do the work I ask of it.

In this chapter, I address Merleau-Ponty's later ontology of "the flesh," through which the mature Merleau-Ponty and some of his contemporary ecophenomenological acolytes think one may address this sort of issue. Turning to the flesh might provide a solution because its radically flattened ontology seemingly permits one to dialectically engage with the world's *own* latent intentionality at every turn. Since, for the later Merleau-Ponty, phenomenological openings are secondary abstractions from a primary ontological unity—a common flesh—from which *all* meaning derives, we're afforded the possibility of questioning and subsequently expressing "what in its silence [the world] *means to say*" (*VI*, 39). The flesh seemingly affords a site through which we might appreciate not only how the world is for us but also the way things are for the world itself.

But there is a serious problem here. Central to Merleau-Ponty's later ontology is his claim that expressive truth consists in the hyperdialectical movement between one's (reflective) articulation and the world's own silent *logos*. Although these poles cannot coincide, their dissonance allegedly constitutes a "good error" (*VI*, 125), which permits true expression because both poles are enveloped within the same flesh. However, even after his ontological turn, Merleau-Ponty emphasizes the requirement for "*hyper-reflection*" (*VI*, 38) to weed out reflective distortions and unwarranted sediment from one's situated account of things. The need for critical reflection suggests that the movement is subject to success conditions, according to which an expression may fall short. To distinguish true (or truer) reflections upon the *more-than*-human world from misleading ones and, indeed, to establish the truth of any underpinning fundamental ontology, Merleau-Ponty must appeal to some measurement of the distance between that reflection and the prereflective sense upon which it draws. However, this measurement, though pivotal to improved claims about what the (rest of the) world "means to say," is, given the necessary interdependence of *logos* and articulation to express truth, rendered impossible by Merleau-Ponty's own lights.

I argue, therefore, that we should be cautious about embracing the flesh ontology and making the sorts of claims that the flesh apparently licenses about the world's intentions as a means of supporting and further unpacking our view-from-everywhere. More specifically, I argue that not only does Merleau-Ponty's later ontological shift fail to provide any substantial improvement on the dialectical position advanced in the *Phenomenology*, but, by resituating the motor of expression within what is purportedly the world's "own" intentionality, appeals to the flesh may shift focus *away* from the very critical self-reflexivity that must underwrite any ecophenomenological attempt to disrupt the violence of our submerged mass.

6.1 THE FLESH

To substantiate these claims, we must first understand what is meant by "the flesh" and the reasons that motivate Merleau-Ponty's turn to it.[1] Like Ted Toadvine (2009), Rudolf Bernet (1993), and Renaud Barbaras (2004), I understand Merleau-Ponty to be consistently engaged in one overarching project: namely, dismantling the distorting binarisms (e.g., subject/object or mind/body) that threaten genuine philosophical analysis. In earlier works, as we've seen, Merleau-Ponty advocated a return to one's phenomenological experience—which problematizes such binarisms—as the necessary ground of philosophy. However, by allegedly beginning "from the 'consciousness'-'object' distinction" (*VI*, 200), Merleau-Ponty came to think that his early work perpetuated some of the reflectively distorted inheritance he sought to challenge. I think it is essential to understand his later flesh ontology as an attempt to address any residual Cartesian diplopia, yet one which is engaged in the same investigation of experience of which science, for instance, provides only a second-order account.

One example of how this motivation plays out is through Merleau-Ponty's mature analysis of the phenomenology of the Other. As we saw in chapter 4, Merleau-Ponty recognizes that our inescapable perceptual faith reveals that, while Others open onto the world distinctly, we each fundamentally experience the same world. Merleau-Ponty came to think, however, that this observation clashes with his earlier "philosophy of consciousness" (*VI*, 141) because alternative openings cannot obviously be *present*ed to consciousness in the requisite (i.e., positive) manner. The Other opens onto the same world and so isn't given as a pure lack, but neither is the Other given as a positivity in the seemingly requisite sense. Resisting an objectivistic or naturalistic ontology as his starting point, Merleau-Ponty previously tried to solve this puzzle by appealing to the shared, nondiscursive anonymity of our experiences. However, Barbaras contends, Merleau-Ponty found himself caught between two stools: either both parties are reduced to the *same* anonymity "in which all alterity finds itself dissolved" and the distinct consciousness to which the Other is presented disappears, or otherwise, before consciousness (as Emmanuel Levinas earlier claimed), "no alter ego can appear" (Barbaras 2004, 38).

Given that such articles of perceptual faith are, Merleau-Ponty says, "beyond proofs" (*VI*, 24), he was led to abandon the purely archaeological focus of his previous work. Crucially, we must remember, this is because philosophy must be amenable to alterity in the sense that it may "tell us how there is openness without the occultation of the world being excluded" (*VI*, 24), something that phenomenology—qua philosophy of consciousness—could apparently never do. Merleau-Ponty's analysis concludes in an ontological shift where subject and object, self and Other aren't foundational terms, but abstractions from a primary unity: a common flesh which sustains their ontological continuity. Through our embodied inherence within the same flesh, both "I" and the "Other," therefore, experience the same "world," and the world may express *itself* through our differential participation within it. The other party's perspective, the later Merleau-Ponty thinks, is thereby made somewhat available to me as a "fold" or "hollow" (*VI*, 146, 196) in our shared flesh, rather than existing as either pure presence (object) or absence (consciousness).

6.1.1 Does Phenomenology *Require* the Flesh?

Before further exploring the flesh, we should emphasize that Merleau-Ponty's motivation for an ontological shift differs subtly from the task I, and other ecophenomenologists, set for it. Merleau-Ponty thinks that, irrespective of any specifically "environmental" benefits, addressing alterity *in principle* requires resituating phenomenology in an ontology along the lines of the flesh. I think we can safely dispute Merleau-Ponty's contention, however, by challenging the claim that phenomenology is *necessarily* a philosophy of consciousness in the requisitely problematic sense.

As Barbaras unpacks the dilemma, phenomenologists either must absorb Others into *their* experiential "anonymity," or cannot really address Others at all. Understood as such, phenomenologists appear bound to the same *absolute*/other dichotomy that Levinas invokes. However, I've previously argued that Levinas mischaracterizes phenomenology by relying on the false dilemma of presence to/absence from consciousness that, as embodied, the Merleau-Pontian body-subject problematizes. If nondiscursive grounds of experience form the basis of the body-subject's anonymous bond with the world, then, *as grounds,* these are never simply and unambiguously manifested at the level of full experiences. Thus, the appeal to common perceptual anonymity needn't entail dissolution of alterity because, at the level of whole perceptual expressions, one is only ever *partially* implicated in the Other's perspective. This is why Sonia Kruks may claim some insight into her colleague's pain without it becoming a manifestation of her own. It's also why Merleau-Ponty may talk meaningfully about a dung beetle's teleological intentions without subsuming the dung beetle's world into his. If I'm right about this, then an ontological turn isn't *required* to appreciate alterity in any obvious sense.

We might also question what Merleau-Ponty means to designate by the term "philosophy of consciousness," which renders such philosophies incompatible with the appreciation of alterity. Merleau-Ponty cannot mean to suggest that we should abandon the irreducibly revelatory nature of experience. Even for the later Merleau-Ponty, "there is certainly no question of abandoning the viewpoint of consciousness" (Barbaras 2004, 128); that is where the all-important experiential alterity is manifested. Indeed, for Merleau-Ponty, the flesh shows itself only (indirectly) through experience (*VI*, 175). His is an ontology allegedly required by experiential consciousness; it is a radicalization of phenomenology, not one which breaks with it. What Merleau-Ponty appears to have in mind is that an unsurpassable barrier to alterity is erected by philosophies which understand consciousness as a Sartrean nothingness or "retreat of not-being" (*PP*, 400), as Merleau-Ponty does in some later passages of the *Phenomenology*. Conceived as such, one *couldn't* register the Other's differential phenomenal opening because, strictly speaking, there would be no-thing to be presented as such.

But this problematic theoretical commitment is one that Merleau-Ponty shouldn't have made in the first place. First, because, as Toadvine (2009, 74–75) notes, Merleau-Ponty thereby reifies a dualistic cleavage by which the "personal self" of reflective consciousness becomes ontologically discontinuous with the unreflective "natural self" of animal life in an exceptionalistic manner similar to that of René Descartes and Quentin Meillassoux. To radically detach motor intentionality from act intentionality would be grossly misleading and at odds with Merleau-Ponty's broader (and more plausible) account. Second, by making perception a Sartrean dialectic between being and nothingness, Merleau-Ponty seemingly neglects his own warning about taking perceivable things to *be* pure positivities. Such Sartrean commitments aren't obviously compatible with the Gestalt ontology with which Merleau-Ponty's account begins, and which provides much of its ecophenomenological merit.

Furthermore, this problematic theoretical commitment is one that Merleau-Ponty *needn't* have made either. Merleau-Ponty had earlier remarked that phenomenology must be "*practised and identified as a style or manner of thinking*" (*PP*, viii) rather than understood in terms of commitment to the axioms of phenomenological texts. Conceived as such, phenomenology cannot *begin* with commitment to Husserlian, Schelerian, or Sartrean accounts of consciousness.[2] So, the question for the radically reflective phenomenologist is whether these philosophies best deal with the phenomena they're employed to explain. Like the later Merleau-Ponty, we can see they do not. Moreover, these philosophies of consciousness are based upon reflective abstractions which unduly exclude or obfuscate the phenomena available for expression. I refer most pointedly to the axiological meanings which cannot be accommodated within the Sartrean schema for much the same reason that they evade natural science: they cannot be straightforwardly identified within the realm of being-*in-itself*. Nevertheless, especially if we should understand phenomenology as primarily an archaeology of experience, I see no reason, independent of further theoretical commitments, why Merleau-Pontians *must* adopt the terms of the Sartrean schema and thus situate phenomenology as a philosophy of consciousness in the problematic sense.

The other issue that troubled Merleau-Ponty was that his earlier phenomenological approach requires "prereflective contact of self with self," or a "tacit cogito," to justify beginning analyses from the viewpoint of consciousness (*VI*, 171). Since, however, any process of reflection requires language—complete with its "sedimented significations"—the "tacit cogito" cannot seemingly be primordial in the requisite nondiscursive sense (*VI*, 171). Consciousness is, for the later Merleau-Ponty, something linguistically or culturally constituted. In his late essay "The Child's Relation with Others," Merleau-Ponty supports this hypothesis empirically. Following Jean Piaget, he argues that developmental psychology shows the unified "I" of consciousness to be derivative of an "anonymous collectivity" or "undifferentiated group life" into which one is purportedly born (*CRO*, 119). One only learns to differentiate oneself from Others (at six or more months of age), Merleau-Ponty claims, by first learning—under the influence of language—to see their bodies *as other*. Thus, according to the later Merleau-Ponty, consciousness and even one's body schema are derived from the primacy of an undifferentiated flesh.[3]

Although, if substantiated, Merleau-Ponty's claims would provide reason to resituate phenomenology within the flesh, we should have reservations about them. Shaun Gallagher and Andrew N. Meltzoff, for instance, draw upon more recent work in cognitive science to argue that children possess, from birth, a "primordial" or "'embryonic' notion of self," and so "*experientially*, and not just objectively, we are born into a world of others" (1996, 226–27). If Merleau-Ponty were right about our primordial state of indivision, they argue, then the "invisible" imitative behaviors Meltzoff observes in newborn children (e.g., where children mirror others' facial expressions, despite being unable to see their own faces) would be straightforwardly impossible, a hypothesis that Merleau-Ponty himself concedes (*CRO*, 116–17; Gallagher and Meltzoff 1996, 220–21). If Gallagher and Meltzoff are right about these basic, nondiscursive, aspects of experience, then we're given some vindication in beginning analyses from a phenomenological viewpoint, because the self of the "tacit cogito" appears to be more than a "solipsistic" abstraction. Furthermore, since Gallagher and Meltzoff argue that even newborns must possess a relatively well-developed body schema to behave as they do, we're given further justification in retaining the sort of phenomenological model explicated in the *Phenomenology*.

More could be said about all of this. If I'm on the right lines, however, then Merleau-Pontian ecophenomenology isn't necessarily incompatible with the balance of occultation and openness necessary for the appreciation of more-than-human alterity, even without any further ontological commitments of the kind Merleau-Ponty posits. The model explicated throughout *most* of the *Phenomenology* appears, therefore, to be at least theoretically compatible with the appreciation of alterity, even without recourse to the flesh. If so, then we can safely return to our guiding ecophenomenological question: Does turning to the flesh facilitate an *improvement* in our ability to address the more-than-human world—in its alterity—over and above our current ecophenomenological model? To answer this question, I shall first attempt to unpack Merleau-Ponty's flesh ontology.

6.1.2 What Is the Flesh?

Contrary to Michel Foucault's (1977) contention, the later Merleau-Ponty never claims that each thing possesses its own self-sufficient meanings, merely awaiting amplification. In the context of the flesh, expressive truth is the culmination of a natural *logos* which calls to, and is called for by, situated (reflective) articulation. Since both poles are bifurcations of an original ontological unity in the context of which they gain their identities, it's erroneous to speak of *my* delivering the world's meanings like an expressive midwife. Each true perception of a more-than-human entity is the "self-palpitation of the world's flesh" (Toadvine 2009, 96): an abstract moment of the movement between poles that expresses the meaning of the whole. By resituating the thrust of expression on the side of the world, the flesh purportedly affords the improved possibility of dialectically addressing more-than-human entities *on their own terms*.

Moreover, the flesh accommodates "identity within difference" (*VI*, 225) because the aim of perception isn't coincidence with some-thing's own, independent, meaning, but the self-expression of fleshy truth via the movement between poles. The Other's alterity is thus retained, because divergent expressions may truly express the same world. This same reasoning means that even nonsubjective things—and the more-than-human world itself—may retain a radical alterity, since their truths aren't exhausted by any expression. As I explain below, a key source of the flesh's environmental promise lies in providing solid ontological grounds through which to retain alterity by refusing to reduce the world's meanings to anthropomorphic or (ethic-guiding) anthropocentric representations-for-us.

Merleau-Ponty's paradigm example of where one may glimpse the flesh comes through the observation that, when I clasp my hands, I may feel my right hand as touching or touched, but never both simultaneously, because the body's imminent self-perception "always miscarries at the last moment" (*VI*, 9). Between "my body touched and my body touching," Merleau-Ponty argues, "there is overlapping or encroachment" (*VI*, 123). The two exist in a *chiasmic* relationship, emblematic of identity-in-difference, because, as Toadvine notes, experience refuses to either assimilate or juxtapose "touching" and "touched" (2009, 113).

Embodiment provides the requisite ontological continuity for chiasmic encroachment: it's only because one's hand may be touched that it may touch, that one may be seen that they may see. "In this sense," Merleau-Ponty argues, we "see and touch the visible, the tangible, from within. Because our flesh lines and even envelops all the visible and tangible things with which it nevertheless is surrounded, the world and I are within one another, and there is no anteriority of the *percipere* to the *percipi*" (*VI*, 123).

The reversibility of our bodies becomes an "*exemplar sensible*" (*VI*, 135) for the purportedly *necessary* intertwinings between self, Other, things, and world. The flesh ontology thereby licenses a profound shift in how we understand our epistemological and ontological relationships. Rather than one's embodiment affording the *possibility* of a dialectic with various aspects of the more-than-human world because of its erosion of a difference in kind between subject and object, sharing a common flesh affords the *same* imminent reversibility experienced between touched and touching hands. Hence, "there is no problem of the *alter ego* because it is not *I* who sees, not *he* who sees, because an anonymous visibility inhabits both of us, a vision in general" (*VI*, 142). The reason for this emerges once we abandon consciousness as the "primordial definition of sensibility" and "rather understand it as the return of the visible upon itself" (*VI*, 142). The flesh thus grounds a "natal bond" with *all* more-than-human entities, which philosophies of consciousness must apparently reflectively misrepresent "by undoing it in order to remake it, only by constituting it, by fabricating it" (*VI*, 32).

Of course, self-other reversibility can only ever be imminent, otherwise one's body would cease to be a boundary of sorts and the other party's alterity would be implausibly lost. Hence, despite radical connaturality between parties, philosophy cannot aim for primitive coincidence between any (reflective) articulation and its underpinning *logos*. True expression is possible only because a fleshy gap ("écart") accommodates the hyperdialectical movement (*VI*, 7, 148). This divergence enables Merleau-Ponty to speak of perception enabling a "good error" which expresses truth because the poles of a chiasm (and the flesh affords several: sensible-sense and nature-culture, for example) are primordially enveloped within the same flesh whose truths unfold *themselves* through the expressive movement.

So, what *is* the flesh? Merleau-Ponty stresses the "elemental" primordiality of the flesh, but it isn't a separate kind of "stuff."[4] This "tissue," he emphasizes, "is not a thing, but a possibility, a latency and a *flesh* of things": a "*general thing*, midway between the spatio-temporal individual and the idea"; "the concrete emblem for a general manner of being" (*VI*, 132–33, 137, 147). Particular visual perceptions, for instance, aren't most fundamentally intentional relationships *between* vision and visible object, but of "visibility in general," or, as Barbaras says, "a single dimension of visibility and, consequently, variants of each other" (*VI*, 149; Barbaras 2004, 201). The flesh, therefore, needn't require us to reify a distinct entity over and above the multitude of possible poles we locate in our worked-over analyses, although these poles exist *as* themselves only through their particular expressive crystallizations.[5]

Since the later Merleau-Ponty reconceives expression as a dialectical or dialogical manifestation of the flesh's autoaffection, the flesh seems to decenter consciousness in the manner necessary to license claims like the following: "One can say that we perceive the things themselves, that we are the world that thinks itself—or that the world is at the heart of our flesh" (*VI*, 136–37). These claims underscore the flesh's apparent ecophenomenological appeal over and above Merleau-Ponty's previous model.

6.2 THE FLESH'S ENVIRONMENTAL POTENTIAL

While the above may permit speaking of what the world "means to say," this shouldn't be taken too narrowly. Isis Brook emphasizes that the sensible-sentient chiasm is a "remarkable variant" of a common flesh, not its sole nature (Brook 2005, 360; *VI*, 136). Despite their appeal in legitimating a straightforwardly reciprocal human-world dialectic, the animist, subjectivizing implications that commentators like David Abram (1996) draw from the flesh, therefore, overgeneralize a particular, remarkable, chiasm.

Merleau-Ponty writes that "when we speak of the flesh of the visible, we do not mean to do anthropology, to describe the world covered over with all our own projections leaving aside what it can be under the human mask" (*VI*, 136). The flesh, Merleau-Ponty stresses, "is not hylozoism: inversely, hylozoism is a conceptualization—A false thematization" (*VI*, 250). As Melissa Clarke and Bryan E. Bannon note, by appropriating the more-than-human world under the narrow terms of human or humanoid experience (i.e., making it the same sort of dialogical *subject* as Karen J. Warren and Carolyn Merchant do), such animism would undermine the very alterity that the flesh was partly inaugurated to explain (Clarke 2002, 216–18; Bannon 2011, 332).

There may, nonetheless, be benefits available due to our primary engagement within the flesh. Brook (2005, 361) suggests that the flesh would dramatically limit the extent to which expression could be anthropomorphized because "nature" (and any meanings therein) couldn't be reduced to our construction of it. This may, in turn, diminish the residual force of any de facto correlationist charge. I've suggested that Merleau-Ponty's earlier phenomenology didn't *entirely* curtail the concern that, as a situated body-subject, one is prone to unwittingly impose the norms of engagement specific to that purview onto nonhuman entities. The doctrine of expression employed in the *Phenomenology* mitigates this problem in two ways: by admitting of multiple true, yet divergent, expressions of the same world; and by recourse to radical reflection (supplemented by intersubjective dialectics), which may bring one's imposition of onto-epistemological violence to the fore. Nevertheless, one is here only *partially* implicated in the rest of the world. The space for anthropomorphic misrepresentation or construction essential to the charge is, therefore, left somewhat open. However, if, in the context of the flesh's *auto*affection, perception becomes "the coiling over of the visible upon the seeing body" (*VI*, 146), then our capacity to address the world on its own terms may improve.

In fact, if *all is flesh* and its poles are inherently *reversible,* as Merleau-Ponty suggests, then his ontological shift *appears* to facilitate something roughly equivalent to an intersubjective dialectic with the more-than-human world. This isn't because nonhuman entities become straightforward dialogical "subjects," but because, according to Merleau-Ponty's radically decentered ontology, any expression of a worldly entity is always already licensed by the flesh from which both parties derive, and which provides their ultimate justification. Given one's privileged access to a fleshy medium through which to communicate, one may, through the dialectical movement of expression, literally question the things themselves and reliably appreciate the answers (i.e., truthfully express the meanings) they provide without subjectivizing them in a misleadingly anthropomorphic sense.

One final benefit is that Merleau-Ponty's later ontology licenses a radical rethinking of the human/nature dualism by resituating the human-nature chiasm in the context of the flesh. Since, as I argued in chapter 2, human/nature hyperseparation licenses the colonial assumptions characteristic of our current naturalistic paradigm, decentering our praxis in this new light offers promise. Although a primordial fleshy "kinship" (*VI*, 133) wouldn't straightforwardly necessitate deep ecological commitments—which risk prioritizing identity over difference in that particular chiasm in a suspiciously dualistic manner—a successful fleshy ontology may at least helpfully motivate rethinking our expressions of these relationships.[6]

6.3 TENSIONS IN THE FLESH

Contrary to appearances, however, the suitably rigorous account of fleshy truth needed to secure these additional benefits seems to necessitate a desituated "soaring over" (*VI*, 111) and thus exceeds the limits of (even a radicalized) phenomenology. Neither the outline of the ontology nor truth claims drawn from it need possess a Cartesian level of certainty. However, to be of merit to our current *eco*phenomenological task, there must be some method or yardstick, over and above the tools available in Merleau-Ponty's earlier work, by which the flesh facilitates improved judgments regarding the expression of more-than-human phenomena. I will now clarify why the flesh cannot grant these and so constitutes no significant improvement on the ecophenomenological model hitherto explored. Doing so requires exploring tensions between alterity and reversibility.

Recall that the environmental promise of the flesh derives largely from its reversibility, which, in our case, extends from the body. In order that expression shouldn't become solipsistic or anthropomorphic, Merleau-Ponty claims that the "body unites us directly with the things through its own ontogenesis, by welding to one another the two outlines of which it is made, its two laps. . . . It is the body and it alone, because it is a two-dimensional being that can bring us to the things themselves, which are themselves not flat beings but beings in depth, inaccessible to a subject that would survey them from above" (*VI*, 136). Merleau-Ponty thus hopes to facilitate expression of the world's truths through our inherence in the same "carnal being" (*VI*, 136) shared with more-than-human Others. Encroachment between fleshy poles allows Merleau-Ponty to claim that, through the hyperdialectic, "the seer and the visible reciprocate one another and we no longer know which sees and which is seen" (*VI*, 139).

One's relationship with the Other is, therefore, purportedly that of a "Sentient in general before a Sensible in general" (*VI*, 142). Since this relationship echoes the experiential reversibility of one's hands—and is supported by our unshakable conviction that we share a common world—Merleau-Ponty argues that "the handshake too is reversible" (*VI*, 142). It is this reversibility, to emphasize, which licenses an alleged improvement in our ability to address the world on its own terms. Nonetheless, if self-other reversibility mirrors the reversibility of one's own hands, it becomes vulnerable to neglecting the other party's alterity by either an excess or a lack.

6.3.1 Missing Others by a Lack

At face value, the handshake reflects a self-Other chiasm under which we share an imminent reversibility because one's experience "does not notably differ from that of the others" (*VI*, 146), presumably because we both partake in "general" relations of sensibility and so on. However, for reasons similar to Fanon's and Young's, Luce Irigaray contends that any such "reversibility" merely naturalizes the terms of one's own experience, resulting in a "labyrinthine solipsism" (1993, 157) inherent to the ontology of the flesh. Recall Merleau-Ponty's claim that "it is not *I* who sees, not *he* who sees, because *an anonymous visibility inhabits both of us, a vision in general*." When contextualized within the flesh, as I unpack it here, such statements seem to take Merleau-Ponty's various claims about anonymity beyond his own experience and desituate them as transcendental truths about *all* (at least human) beings.[7] Thus, the problem of the *alter* ego is erased, Irigaray thinks, but only by collapsing differences into the problematic "masculine" anonymity described, at times, in the *Phenomenology*.[8]

The experiences of two human beings sharing very similar bodies and social norms aren't straightforwardly reversible. Given that we've already ruled out their animate subjectivization, the pregnant silences of nonhuman entities are surely further beyond our reach. If *my* account of what the *world* (or, perhaps, the Arctic tundra or "depleted" rainforest) differentially wishes to say in heterogenous contexts relies upon an implausible reversibility, then any improved possibility of expressing rather than obfuscating its meanings is stifled, and Merleau-Ponty's claim that the two poles share a fundamental identity is unwarranted. This would be devastating for the claim that the flesh better permits one to address the world's truths on their own terms.

Moreover, taking Irigaray's arguments seriously, even the sexuate markings of one's human body, for example, may (come to) shape one's account of intersubjective participation in, and of, any flesh. Since one's situated embodiment affords only a partial opening onto the world and, ultimately, any flesh, therefore, one's *fundamental ontological account of the flesh* will presumably itself be similarly skewed by one's limited mode of participation within it. It's difficult, therefore, to see how the environmental benefits previously outlined remain plausible if one's experience and ontological contentions remain dominated by anthropomorphic (and andromorphic) projection.

But there is a certain irony in this revelation, given that it was this very concern about ecophenomenology's propensity to subsume more-than-human entities into an androcentric or ontologically anthropocentric horizon that motivated our investigation of the flesh's capacity to provide a proxy for intersubjective dialectic in the first place. If turning to the "general" relations of the flesh encourages us to overwrite alterity as Irigaray suggests, however, then such an ontological shift would constitute a *reduction* in openness to a genuinely *more-than*-human world. In Irigaray's reading, coinstantiation in the flesh appears to license the same problematic "shortcut" to the other party that Shannon Sullivan links to colonial violence. For Irigaray, the flesh merely ontologizes and, thus, naturalizes the hegemony to which phenomenology is already prone.

Of course, Merleau-Ponty never claims that I can flip from my experience of another's sensibility to experiencing *her* sentience, despite their fundamental "generality." He's aware that to do so would be to undermine alterity and the movement by which meanings come to fruition in the flesh. Bannon (2011, 332–33), therefore, moots one response (i.e., Clarke's) to our quandary: to concede a fundamental *asymmetry* in reversibility. As Bannon notes, however, this will help little. Asymmetry would ensure that the meanings one locates in the world remain vulnerable to the charge of anthropomorphism which the flesh—in its apparent capacity to facilitate radical human-world reciprocity—is employed by ecophenomenologists (like Brook) to curtail.

Moreover, Clarke (2002) seems to think that—as a sensual relationship in which they cannot partake—reversibility cannot extend to inanimate things. An asymmetric reversibility may be permitted between us humans and the perspectives of the axolotls or hyraxes we seek to accommodate in a view-from-everywhere. However, there can be no reversibility whatsoever regarding the meanings (or alleged nondiscursive grounds thereof) I might claim to find in retreating glaciers or rising sea levels. In either case, the flesh ontology *itself* facilitates no obvious improvement in terms of one's ability, as a situated human body-subject, to reflectively uncover those rich meanings, nor to disrupt one's imposition of onto-epistemological violence. By following Clarke's lead, alterity apparently comes at the cost of openness, yet—as our exploration of absolute alterity demonstrates—*both* are necessary for ecophenomenologists to secure the reduction in violence that the flesh promises over and above Merleau-Ponty's previous model.

Bannon argues, therefore, that if we accept that the essential epistemological and ontological continuity between parties is achieved *only* through "the projection of a perceptual structure found within human (though very likely all sentient) experience" (2011, 333), then both our *fundamental ontological account of the flesh* and *any* meanings licensed by it will reflect anthropomorphic, or at least biomorphic, projection.[9] If the poles of the flesh—including what the flesh "wishes to say about itself"—*lack* its characteristic reversibility, however, then Merleau-Ponty's ontological turn licenses little increased hope of truly expressing the world's own "*latent* intentionality" (*VI*, 244). Our new ontological context may even *worsen* our ability to accommodate nonhuman alterity, because the fundamental grounds of any "reciprocal" human-world dialectic are, here, both unavoidably asymmetrical and apt to conceal that asymmetry under the guise of the flesh's autoaffection.

But maybe Irigaray straightforwardly misunderstands fleshy reversibility. Clarke is certainly right about imminent reversibility being incompletable, not least because the divergence between poles is essential to express truth. Thus, since the *pure reversibility* implied by Irigaray's critique would close this (fleshy) gap, it should be ruled out a priori.[10] Merleau-Ponty evidently means to say more than that one's experience of the Other's sentience, for example, simply happens to miscarry at the last moment. Pure reversibility would be straightforwardly inconsistent with Merleau-Ponty's account of the alterity involved in divergent expressive openings onto the more-than-human world. Moreover, pure reversibility would seemingly preclude access to nonsentient entities which still maintain latent meanings (or grounds thereof) to be expressed. Since the problem of alterity motivates Merleau-Ponty's ontological shift and fuels much of its corresponding ecophenomenological appeal, I think that we can rule out the pure reversibility interpretation.

A more plausible interpretation is informed by Merleau-Ponty's contention that poles are *individually incomplete* codeterminants in the true expression of meaning. According to this interpretation, "reversibility" refers to the world and body-subject's coexpression of truths, which are true *in*, and *because of*, the divergence between them. Since, however, each pole only becomes itself through the expressive movement, neither is anterior to the other and expressions *are* as much one party's as the other's. This interpretation helps explain Merleau-Ponty's contention that "each is only the rejoinder of the other, and . . . therefore form a couple, a couple more real than either of them" (*VI*, 139). "Expression," here, doesn't "refer to a perspective *upon* the world, a perspective that would be distinguished from the world; expression is the world itself, *as* the world putting itself into perspective" (Barbaras 2004, 233). Reversibility needn't entail, however, that mine is the only sort of perspective through which the world may express itself. Chiasmic logic, suitably resituated, may therefore provide the means by which to resist phenomenology's residual anthropomorphism.

The central problem with critiques like Irigaray's, the argument goes, is that they try to incorporate the doctrine of reversibility at the level of situated consciousness which, by assimilating poles, inevitably undermines the Other's alterity. It's more plausible (and largely in keeping with Merleau-Ponty's final text) that reversibility applies to the fundamental, fleshy level and doesn't straightforwardly translate to subsequent levels in which we abstract consciousness and object. This account of reversibility may still problematize the other party's *absolute* alterity by beginning from *beneath* a philosophy of consciousness. *Reversibility in the flesh* may thereby facilitate an improved capacity to recognize encroachments between parties through a unitary ontological heritage that licenses some grasp of the "general" relations we may share, explaining our primordial faith that we experience a common world, albeit heterogeneously.[11] Merleau-Ponty, however, repeatedly emphasizes "overlapping and fission, identity and difference" (*VI*, 142). Reversibility so understood needn't demand reduction to the same.

6.3.2 Missing Others by Excess

If reversibility applies only at the fundamental fleshy level, then genuine alterity may be retained at subsequent levels of investigation. A very strong doctrine of identity-in-difference is also accommodated because even "incompossible" expressions may be simultaneously true. In this interpretation, for instance, it would be erroneous to claim that the world wasn't (as

Merleau-Ponty [*PP*, 187] had earlier characterized expression) "singing" *itself* through Merleau-Ponty's perceptual experiences and theoretical contentions simply because they cannot be reconciled with Fanon's or Young's. The world's silent *logos* can express itself divergently without either exhausting itself or undermining the dialectical reversibility which is pivotal to the flesh's ecophenomenological appeal.

The danger in our alternative characterization, however, is that it's difficult to see how such a strong commitment to identity-in-difference wouldn't cause one to miss the other party by an excess because, *if all is flesh*, almost *all* expressions seem to become true merely by virtue of being expressed. Our alternative interpretation would thereby risk undermining the alterity of more-than-human entities, since almost any expression of "their" truths would be true by default, regardless of its content. By purporting to maintain such a strong claim to alterity, our alternative interpretation seems to naturalize even *radically* anthropocentric expressions through the logic of the "good error." If all is flesh, the only straightforwardly "bad errors" are those that *entirely* overwrite prereflective alterity. It's not clear that even radically anthropocentric expressions (in even the most troubling ontological and ethical senses) must do so unless they maintain an exclusive and exhaustive claim to truth. Thus, although not exclusively so, certain viewpoints that environmental philosophers aim to challenge by appealing to the flesh, such as the "masculine" perceptual frameworks that underpin technoscientific exploitation, for example, appear to be justified by this model.[12] Furthermore, if exploitative masculinist expressions are *no less true or revealing* than others, it isn't clear that the flesh must heighten, or even retain the impetus for, Sally Fischer's call to heed alternative—and, one wants to say, more warranted—viewpoints to facilitate better communal strategies of environmental engagement.

It gets worse. Since all is flesh, the flesh purportedly licenses a further chiasm between the "silent" world of perception and the linguistic world of *reflective* expression. Thus, although philosophical reflections are historically and culturally situated—and necessarily so, since reflections are only thinkable within one's culturally contingent language(s)—the noncoincidence between poles constitutes another "good error." The "error" is, again, "good" because the thinking body-subject is implicated in "carnal being" through her fleshy embodiment, so there exist the requisite relationships of encroachment and divergence for the world's own truths to be hyperdialectically expressed (*VI*, 120). The nature-culture chiasm is significant here, since one's discursive or culturally delineated reflections manifest what the world "means to say," apparently constituting a "screen" (*VI*, 150) without which the world's *logos* couldn't be shown.

If the logic of the "good error" makes virtually all expressions true, then our problem from what Barbaras calls the dimension of "sensibility" is replicated within the dimension of "intelligibility" (2004, 304–5). A very strong principle of identity-in-difference would allow almost all reflective abstractions—even neo-Cartesian contentions about "inert" or "worthless" nature—to be *just as true* as those contained within Merleau-Ponty's final texts.[13] Since the appeal of Merleau-Pontian ecophenomenology—and the flesh in particular—lies in contributing to more positive and critically self-reflexive relationships with(in) the more-than-human world, it is markedly damaged if it ends up legitimating almost all perceptions of, *and* reflections on, the world as genuinely reversible expressions of *its logos*.

It's tempting to think that this issue isn't perhaps as serious as I suggest. I've previously problematized any correspondence theory of truth which would demand that the aforementioned neo-Cartesian judgments coincide with the world's "essential" properties. Moreover, I've suggested that *some* reductive scientific expressions, where well-informed, *might* constitute regionally true models of the world as constituted by "natural resources," "carbon deposits," and so on. When reconfigured within the context of expressive truth, their theoretical apparatuses may *limit* rather than straightforwardly misrepresent their target phenomena, and so express *something* true of the world's meanings, even if they cannot *themselves* admit the richer range of meanings available there.

Nonetheless, we must remember that we're exploring whether, in lieu of a straightforward intersubjective dialectic, fleshy "reversibility" might provide *improved* means of grasping the more-than-human world on its own terms. But if *all* expressions somehow constitute the flesh's autoaffection, then, in most cases, it's difficult to see how there could be "better" or "worse" expressions of nature's *logos* at all, let alone how the flesh ontology itself could make us *more* critically (self-)reflexive about the meanings we locate. Moreover, if the flesh's self-expression also requires movement between intelligibility and sensibility, then we appear to sacrifice the fruitful critical tools licensed by understanding scientific expressions like Paul Crutzen's to be second-order abstractions. If the flesh requires reflective expression for its meanings to come to fruition, then we cannot grant any sort of deliberative priority to phenomenological experience, nor can we grant nondiscursive grounds of experience priority over discourse. Both are just different moments of the same movement.

The central issue here is that the expressive movement by which truths apparently come to fruition in the flesh isn't obviously (self-)*critical* in the requisite manner to license something even roughly equivalent to the intersubjective hyperdialectic. Barbaras claims that Merleau-Ponty's later "philosophy of expression brings forth a *regressive* approach that, starting from culture and language, interrogates their originary soil" (2004, 59). Since the world motivates each expression, yet resists exhaustion, the hyperdialectic is incompletable. The hyperdialectical movement cannot even be thematized, Merleau-Ponty thinks, without being reduced to a distorting reflective philosophy in which poles are reified. Barbaras thus claims that "what defines interrogation, as pure interrogation, is that the question accepts no answer that would conclude it" (2004, 141). In one sense, we can accept Barbaras's claim since we don't seek reflective coincidence between, or reification of, dialectical poles as ontologically primitive terms. The impossibility of either route, however, suggests that the flesh itself could *never* afford substantive additional guidance by which one expression could be thought to be *better* than another. This is problematic insofar as, in the present ecophenomenological context, the flesh is charged with providing *improved* means of facilitating critical (self-)reflexivity over and above those already licensed by radical reflection. By reconfiguring expression as the flesh's autoaffection, however, the hyperdialectical movement appears, *at best*, to call for *more*, not *better*, expressions of "natural *logos*," which is an odd (and, in this context, problematic) use of the term "interrogation."

6.3.3 The Yardstick Problem

Thankfully, Merleau-Ponty doesn't subscribe to this "all true" hypothesis. He thinks (wrongly, in my view) that any rigorous and coherent account of truth is conditional upon one's inherence in the flesh. By bifurcating subject and object, Merleau-Ponty thinks, philosophies of consciousness invert critical reflection such that we may only return to *our own* perceptions for truth conditions by which to judge expressions. Even under Merleau-Ponty's earlier phenomenology—which problematizes the subject/object binary yet makes experience the ultimate court of appeal—the ways things appear to some degree reflects how one *happens* to explore them. Expressions (or our identification of the meanings therein) are thus prone to be anthropocentrically (in at least the risky ontological sense), anthropomorphically or, at a push, biomorphically biased in a manner similar to the one that concerns Bannon.

The flesh permits an escape from this reflective cul-de-sac, Merleau-Ponty thinks, because one's conviction about something's truth is underpinned by its dependence upon a fundamental fleshy reality. One's perceptual expectations are satisfied by the real because it *is* real; things don't become real by being expected (*VI*, 40). Likewise, expressions of the meanings allegedly expressed or made available by more-than-human entities satisfy us because they're true; they don't become true by satisfying us. Something similar was, of course, true of Merleau-Ponty's previous work, where, by contributing to our expressions, the more-than-human world constrains the "real." Merleau-Ponty is unfair in suggesting otherwise. Nevertheless, the flesh is meant to provide further justification for one's critical judgments because one's grasp of "reality" derives from a primary unity between poles.

Toadvine suggests that the flesh, therefore, limits which of our meaningful expressions could be thought to be successful (2009, 127–30). Within the flesh, he claims, there exist indicators of reality upon which to draw because it is within the flesh that one's true vision—as an instance of "visibility in general" which somehow predates the subject-object chiasm and so isn't subject to the same concerns about anthropomorphism—for example, inheres.[14] So, what are these indicators? Toadvine suggests that, for the later Merleau-Ponty, an expression should be judged with respect to its relationship to the *logos* it hopes to express. Toadvine points us to a working note in which Merleau-Ponty speaks of the world's "silent persuasion" (*VI*, 214), which true expressions may voice. How, then, to determine whether an expression (including those which make marginal reference to the meanings located within genuinely *Other* viewpoints) passes this test? We can rule out two routes. One cannot look for expressive coincidence with any "flesh-in-itself" without misunderstanding Merleau-Ponty's later ontology. One also cannot look for coincidence between expression and any "meaning-in-itself." Recalling that "what the world means to say" becomes itself only hyperdialectically, one cannot prize the "cultural" from the "natural" or "primordially given" without misunderstanding their coconstitutive roles.

Toadvine, therefore, highlights another working note in which Merleau-Ponty claims that "language realizes, by breaking the silence, what the silence wished and did not obtain" (*VI*, 176). What Merleau-Ponty seems to have in mind is that, by enacting reflective expression (which is uniquely linguistic and may "break the silence"), it just becomes clear to what extent one has voiced the world's latent intentionality through one's hyperdialectical engagement with it. Just as the world may only "sing" through an expressive act, we experience the subsequent reconversion of that linguistic expression into an unexhausted natural event, itself to be reflected upon.

Neither Merleau-Ponty's unfinished work nor Toadvine's gloss, however, explains *how* the flesh might provide better means of identifying discrepancies between reflective expression and the more-than-human meanings it hopes to voice. Remember (especially given our sensitivity to the creative/derivative expression distinction), we're not asking for definitive or infallible truth conditions, but rather *something* the flesh may provide, above and beyond Merleau-Ponty's earlier model of radical reflection, to facilitate improved means of distinguishing better from worse expressions, thereby informing the view-from-everywhere. However, granted that the interrelational identities of fleshy poles rule out the identification of any reflective target outside the expression itself, this "something" couldn't be given *in principle*. Put differently: in lieu of its capacity to provide something like a yardstick, it's difficult to see how the flesh *itself* affords improved means of evading the egomorphic, andromorphic, anthropomorphic, or biomorphic ventriloquism to which we are prone. Although the later Merleau-Ponty considers the task of philosophy to be the hyperdialectical "reconversion of silence and speech into one another" (*VI*, 129), given the impossibility of its thematization, the yardstick seemingly required to guide reflection to better express, rather than obscure, the (rest of the) world's dialectical interjections and, therefore, realize the potential specific to the flesh, remains necessarily absent.

Moreover, as Bannon (2011, 335–36) argues, if, for the mature Merleau-Ponty, conceptually laden human language is the *only* means by which to give nature "its" voice, then our problems of anthropomorphism and ontological anthropocentrism may increase in magnitude.[15] This isn't simply because of the threat of undisclosed sediment within human language, but because Merleau-Ponty also thereby affords human consciousnesses a certain *onto-theological* priority in revealing the more-than-human world's truths.[16] Such a view threatens the scope and putative reversibility of any dialectic because more-than-human entities—as supposed interlocutors or cocontributors to the view-from-everywhere—may reveal themselves *only* through the asymmetric terms of human expression.

Situating interrogation within this purview may also do little to counter our tendencies toward ontological colonialism if the world may express itself only through *reflective* terms that are apt to distort, rather than refine, our target phenomena (or grounds thereof). It was, after all, Merleau-Ponty's rejection of the reified (e.g., subject/object) terms of reflective philosophies which motivated his return to phenomenological methodology in the first place. However, if, within the context of the flesh, these terms become the culmination of the world's silent *logos* rather than a misleading reflective abstraction, one's focus seemingly shifts further away from the critical self-reflexivity that is pivotal to Merleau-Pontian ecophenomenology's appeal. Of course, Merleau-Ponty needn't necessarily hold that *every* reflective utterance accurately expresses the world's truths.[17] But, even so, the flesh *itself* provides no obvious further means to guide our judgments about them.

Our only real clue to a possible solution comes through what Merleau-Ponty calls "*hyper-reflection.*" Since reflection also "makes use of powers obscure to me" in its "distinct act of recovery," hyperreflection asks us to suspend our ordinary faith in the world to "take itself and the changes it introduces into the spectacle into account" (*VI*, 38). Remaining sensitive to our tendency to reflectively prejudice what we find in experience through the interrogative terms we employ, the later Merleau-Ponty advocates hyperreflection to uncover the intentional debt we owe the world and, somehow, thereby illuminate the meanings ecophenomenologists seek. Hyperreflection purportedly takes place *beyond* the sensible-intelligible conversion and helps one progressively appreciate what was true from one's natural attitude by "plunging" into the world experienced, rather than attempting a mistaken transcendental reflection. Hyperreflection, Merleau-Ponty thinks, provides improved apparatuses with which to make gradual moves toward our "prelogical bond" (*VI*, 38) and, ultimately, to replace misleading expressions with better ones.

The example Merleau-Ponty uses to illustrate hyperreflection is a familiar one: where a "rock" breaks up, on closer inspection, to become a piece of wood. Although both perceptions are initially "real" and only subsequently shattered, they are, Merleau-Ponty thinks, undeniably progressive approximations or manifestations of the same world and the same, familiar, flesh. We should emphasize, however, that hyperreflection is meant to provide further *justificatory* grounds to think that one's grasp improves in this scenario, over and above better satisfaction of one's expectations. Despite Merleau-Ponty's misleading claim that the hyperdialectic makes expressions "all true" (*VI*, 41), hyperreflection, therefore, purports to afford a *method* (if not a criterion) by which to understand expressions to be better or worse than one another in virtue of their fidelity to the (otherwise incomplete) natural *logos* they seek to differentially express.

The glaring issue here, however, isn't only that we're necessarily left without additional success conditions by which to hyperreflect in earnest, but that hyperreflectively *living* the hyperdialectical movement from *within*—which is, evidently, what Merleau-Ponty advocates, given the impossibility of its thematization—must be pursued as a situated, *human,* body-subject. This requirement yields several interrelated issues, each of which is underpinned by the familiar tension between reversibility and alterity.

Given that perception and reflection are not genuinely independent of one another, Merleau-Ponty emphasizes, in both periods, the importance of being alive to the intentional commitments and habitual sediment which shapes them. Thus, in the *Phenomenology,* Merleau-Ponty prescribed radical reflection to uncover a similar debt of meaning to the one identified in his final text. Radical reflection also serves to identify where *unwarranted* sediment shapes perception or reflection in a manner that obscures, rather than clarifies, the world's heterogeneous meanings. Hyperreflection, therefore, plays an almost identical role in Merleau-Ponty's mature philosophy. Since, however, no reflections can be self-sufficient (this would require an implausible act of transcendence), even hyperreflective expressions are likely to be riddled with distorting sediment. In one sense, Merleau-Ponty can admit this without issue since he still takes philosophy to be hyperdialectical. Hegel remains the museum. Nevertheless, while Merleau-Ponty recognizes that (asymmetrical) expressions must be undertaken from within a situation, he doesn't fully appreciate that doing so without the possibility of further guidance beyond that already supplied by radical reflection means that—in lieu of a plausible substitute for robust intersubjective dialectic (which pure reversibility, although untenable, would provide)—the flesh itself puts one in *no better position* to appreciate where the world's differential meanings are obscured or more accurately expressed. *Our* turn to the flesh is thus dealt a serious blow, since its appeal lies in providing ontological grounds that afford improved access to the meanings licensed or expressed by more-than-human entities *on their own terms.*

Some of Merleau-Ponty's examples illustrate this tension between fleshy reversibility and an alleged improvement in openness to alterity. In his late essay "The Philosopher and His Shadow," Merleau-Ponty disavows any philosophy of history that either situates itself "objectively" as self-contained theory (which would "rob the great philosophers of what they have given others to think about") or becomes interpretative ventriloquism ("a meditation disguised as a dialogue, in which we would ask the questions and give the answers" [*PS*, 159]). Merleau-Ponty instead emphasizes the importance of voicing the "unthought" in the work analyzed, whereby "it is indeed in the contact going from one to the other that both become themselves" (Barbaras 2004, 69). The problem here is that Merleau-Ponty was talking about Husserl, whose thought Merleau-Ponty often wildly *mis*interprets by assimilating Husserl's position into his. However, through Merleau-Ponty's later account of expressive truth, there is no substantive mechanism to which one may appeal to justify this judgment, since each philosophy—like the world's latent intentionality—only "truly" becomes itself through the expressive movement.

One major issue in how this plays out is that, beyond the extreme limit cases where something *must* be incompatible with the flesh ontology, we're left wanting. As noted above, *if* necessarily exhaustive, certain Cartesian metaphysical observations would be ruled out because they neglect the alterity of a world left unexhausted by a particular kind of reflective inquiry. However, in my understanding, these kinds of observation were already undermined by radical reflection because of inassimilable tensions between their totalizing assumptions and an existential phenomenological methodology. Turning to the flesh provides no obvious benefit on this score.

Moreover, recalling that, for the later Merleau-Ponty, expression involves a movement between various poles—*each of which inheres equally in the flesh*—one cannot favor any in discerning what the world means to say, even where this involves the nature-culture chiasm. This is problematic in the present context because our rejection of certain objectivistic, dualistic, utilitarian, and technoscientific contentions, for instance, rests to some degree on attributing a certain primacy to nondiscursive grounds of experience that proponents of the flesh ontology cannot obviously admit.[18] Thus, where expressions are neither exhaustive nor otherwise straightforwardly incompatible with the flesh's *basic* ontological stipulations, although some expressions evidently appear more warranted than others, hyperreflective "successes" may be only intuitively bestowed. Intuitions often vary wildly and yet, in the context of the flesh, which admits of no substantive distinction between discourse and the nondiscursive, one has recourse to little further. The most we can require is that one *thinks* one is being hyperreflective in one's judgments. But given our worries about reflective distortion and clandestine sediment, this requirement is surely insufficiently robust to balance alterity and reversibility in the requisite manner to facilitate an *improvement* in the claims that ecophenomenologists might want to make about the more-than-human world's meanings.[19]

Ultimately, in matters of hyperreflection, one may be in an even *worse* position than when one radically reflects in Merleau-Ponty's earlier philosophy. Although the success conditions for radical reflection may be little clearer, its official task is to derive conclusions about primordially coconstituted human-world intentionality. Intersubjective dialogue will certainly help clarify common nondiscursive grounds of experience shared by radically reflective body-subjects and expose unhelpful sediment and reflective distortions in one's guiding view-from-everywhere. As Irigaray's, Young's, and Fanon's criticisms show, however, radical reflection proves difficult. Merleau-Ponty's later task gets even harder. By situating truth in the context of the flesh, hyperreflection calls on one to interrogate whether one's reflective expressions best voice the more-than-human world's *own intentions*—a world with which one is purportedly identified at an elemental level, but which may only be addressed as a situated, human, body, inescapably entangled with the (often problematic) norms and sediment of that situation. The above also exposes two further hurdles specific to the apparent benefit of appealing to the flesh ontology, specifically, to aid hyperreflection. First, the doctrine of fleshy reversibility *cannot* entail pure reversibility and will, therefore, little mitigate the impact of sedimentation or reflective distortion at the level of hyperreflection. Second, no appeal can be made to any substantive distinction in priority between discourse and the nondiscursive.

Thus, it seems that, in hyperreflection, a few extreme cases notwithstanding, one is brought back to one's own concrete situation to make judgments about the meaningful content of more-than-human reality, ultimately (in lieu of a robust yardstick, infallible acquaintance with the world's truths, or a genuinely analogous equivalent to intersubjective dialectic) by whether one happens to think one's reflections better express the world's latent intentionality. One is cast back upon one's own perceptual faith and since, as we've seen Jill W. Scott argue, perceptual faith needn't be a reliable guide to reality, these situated judgments are surely insufficiently robust to constitute any *improvement* in the relationship between reversibility and alterity for the task in hand. Irigaray's criticisms are thus reborn in a new guise: even in the context of the flesh, one's judgments will be skewed by one's limited participation in the world that one hopes to voice, and one's account of hyperreflective "progress" must be established within this same remit. The potential environmental benefits surveyed above are, then, severely diminished since they rest upon the plausibility of the flesh *helping* one better uncover worldly meanings on their own terms in accordance with the view-from-everywhere's improvement curve.

There is also a final tension in employing an ontology of the flesh. If one is carnally implicated in the flesh, one cannot plausibly survey it from outside to make *fundamental* ontological judgments. By making such judgments from within one's situation, however, one cannot plausibly make sufficiently robust, *totalizing*, judgments about the fundamental nature of things that (necessarily) differ so radically from oneself.[20] Nowhere is this tension more keenly felt than where one hopes to advance a fundamental ontology: a theoretical account of *the* ontology of the flesh, which exhaustively describes the preexperiential ontological situation. If our judgments are legitimated only by the flesh, a fundamental ontology would also have to bypass the (sedimented) -morphisms and -centrisms discussed above, which is implausible. Thus, we should be wary, as critical environmental philosophers, of taking up such a fundamental ontology in the first place.

6.4 A WORD ON ONTOLOGY

I've argued that the flesh doesn't license understanding the human-world relationship as any straightforwardly reversible or more reciprocal exchange between equally dialectical "interlocutors." Given expression's fundamental asymmetry and the impossibility of a fleshy yardstick to guide it, hyper-reflection is no less introverted than radical reflection. Indeed, by refusing some of Merleau-Ponty's earlier tools (e.g., a substantive distinction in [non] discursive grounds of experience), it probably fares worse. Appealing to the flesh, therefore, facilitates no obvious ecophenomenological improvement to the hyperdialectic which characterizes Merleau-Ponty's earlier approach.

Some commentators suggest (albeit not always consistently), however, that, although it cannot itself directly authorize specific meanings, resituating reflection within the flesh is useful in instilling a certain mindfulness or sensibility that a profound rethinking of the human-nature relationship licenses.[21] Thus, regardless of any supposed capacity to license an equivalent dialectic to that shared intersubjectively with other human body-subjects, the *cast of mind* the flesh encourages is a distinctly positive influence upon our quest for less problematically ontologically and ethically anthropocentric reflections upon the more-than-human world. I end this chapter by addressing in more detail a counterpoint to this claim which I touched upon above: a move to the flesh may not improve the nonanthropocentrism of our ecophenomenological praxis because, in practice, it typically either provides no clear benefit or encourages a shift in focus *away* from critical self-reflexivity.

For context: the supposed benefit cannot arise simply from the connaturality of parties in the flesh. Even for the earlier Merleau-Ponty, as embodied, one is radically ontologically continuous with the rest of the world. It must instead be down to the *specific* way the flesh flattens ontology, whereby connaturality is radicalized such that human expression becomes partly constitutive of the flesh's autoaffection. I've argued, however, that reconceiving expression in this way will likely obstruct a critically self-reflexive mindset. This is primarily because, by resituating the motor of expression fundamentally *elsewhere* and failing to provide any further identifiable means with which to keep our tendencies toward onto-epistemological imperialism in check, we risk becoming *more* prone to misconceive misleading sediment or our own intentional commitments for the world's differential self-expression.

Let's consider a few diverse examples drawn from the ecophenomenological literature to flesh out this charge. Both Bryan E. Bannon and Monika Langer (1990) invoke the flesh to call for an egalitarian, deep ecological ethic under which we seek to contribute positively to "nature," conceived as a heterogeneous fleshy whole. For Bannon, the flesh may permit "human dwelling" only where it "fosters conditions under which other forms of life (or geography) can flourish and evolve within their own temporal timescale" (2011, 354). But theirs isn't the only avenue that the flesh is taken to motivate. Gary Madison thinks that coinstantiation in the flesh means embracing "postmodernism" and "globalization"; Sally Fischer, that it suggests founding direct democratic institutions; for Molly Hadley Jensen, Merleau-Ponty's "concrete guidance for ecological ethics" is that our primary concern should be to retain species biodiversity; and for David Abram, the flesh even allows one to taste the berry in a blackbird's beak (Madison 2001, 179–80; Fischer 2007, 210; Jensen 2007, 198–200; Abram 1996, 126).

We should pause to note that Bannon's contentions are complicated by the fact that he distances himself from the predominant "monistic" (Bannon 2014, 109) interpretation, considered above, whereby the flesh somehow *itself* licenses its resultant expressions. Bannon emphasizes the Whiteheadian resonances in Merleau-Ponty's later texts to pursue a process account of the flesh, according to which, "ontologically speaking, the flesh of the world is the fabric of space, time, and movement within which we dwell, produced by the interrelation of the myriad bodies that exist. It is not, however, a substance out of which beings are composed, since space, time, and movement are themselves relational processes" (Bannon 2011, 349–50). Within this context, a being just is "the nexus of relations that affect it through the dimensions of its body" (2011, 352), linked through a wider system of contingent relations in their becoming. Since Bannon generalizes the notion of fleshy affect to include *any* aspect of "*a body's* relational engagement with its milieu," he concurs with Toadvine that the flesh *cannot* license any specific moral guidance (2011, 330, 353). His recommendations, to reiterate, don't flow straightforwardly from the flesh ontology; they're intended to show that one *can* develop a normative position mindful of one's intrinsically relational nature (Bannon 2014, 164).

Bannon's suggestions, like much of Merleau-Ponty's work, are really rooted in his desire to erode traditional normative oppositions between "humans" and "nature." His more specific recommendations derive from the impetus not to master the more-than-human world and an openness to the importance of plural more-than-human temporalities or the value of relative "wildness" that *may* (but needn't necessarily) emerge where we're more open to our ontological continuity with (the rest of) "nature." More generally, Bannon *thinks* the flesh to be useful in helping develop our sensitivity to the intertwining interests of various more-than-human parties and how our multifaceted relationships are shaped by our various actions (2014, 161). However, as I've argued, we don't obviously *need* the flesh ontology, monistic or otherwise, to develop this ordinary kind of hermeneutic suspicion or critical self-reflexivity about the meanings or judgments we invoke or extend, or the behaviors to which we commit. Nor is our situation within the flesh ontology itself obviously particularly helpful in developing a sensibility toward doing so over and above its capacity to erode the self/other and human/nature dualisms, which (especially when supplemented by ecofeminist and new materialist insights) was already available in Merleau-Ponty's earlier framework.[22]

As for the others, who do seemingly subscribe to the monistic interpretation, my point isn't that there's nothing of merit in these diverse and often curious accounts. I want to focus on the fact that each thinks that *theirs* truly reflects "what the world means to say" and employs the flesh ontology in support of their cause. Yet, as I've argued, beyond the requirement for a "hyperreflective" mindset, neither I, nor they, appear to have any real means of saying otherwise.

In each of the above examples, this limitation is particularly interesting because the flesh appears to have been selectively employed to support exactly the sort of position that thinker *sets out* to advance in the first place. They each take up the flesh ontology primarily because they (no doubt genuinely) think it speaks to their respective causes or antecedent commitments, and the flesh is intended to provide an extra layer of *support* for them. While I have no in-principle objection to this general sort of move, however, we should be reminded of two dangers in employing the flesh ontology, specifically, to do this sort of work: first, through the ontological flattening it employs, the flesh may license a reduced focus upon critical *self*-reflexivity; second, the flesh ontology itself provides no further usable tools with which to address our imposition of conceptual violence. Thus, the flesh seemingly constitutes a worse (or, at least, no better) framework in which to develop the requisite cast of mind for better engagement with the more-than-human world than Merleau-Ponty's previous model.

Abram (1996, 175–76) provides an exemplary case in point when he discusses the Dreamtime stories of indigenous Aboriginal peoples as definitive instances where the land itself provides ethical imperatives of the sort that human beings, *estranged from our common flesh*, have regrettably *become* deaf to. As Steven Vogel (2006, 157) notes, however, even if specific imperatives are somehow "issued" by landscapes rather than dialogical human interlocutors, at least in the context of the flesh, the question of the accuracy of our translation of them must remain significantly unanswered because that landscape cannot speak to us—*even in reply*—except through *our* expressions of it.[23] Nonetheless, in Abram's case at least, it's partly via appeal to the reversibility of the flesh that the aforementioned imperatives come to derive their reflective justification.

Of course, this fact alone doesn't mean that theirs *isn't* a true expression of the flesh. As Merleau-Ponty himself contends, Abram, Madison, or Jensen might plausibly "grasp" their reflective conclusions (and corresponding resonances in Merleau-Ponty's later ontology) *because* they are true. But, in any case, appealing to the flesh in this manner little encourages the all-important habitus of critique. Although, as I've argued, one is always prone to (mis)characterize the more-than-human world according to one's own intentional commitments (which, as Abram's animism shows, will likely include *at least* problematically anthropomorphic ones), given the inherently reversible way the flesh ontology is usually set up—where one *begins* with the prejudice that one probably has the commitments one does because they are extensions of the flesh's autoaffection—one appears to become *more vulnerable* to overlooking this problem than under Merleau-Ponty's previous model. This might not prove wholly disastrous when we're dealing with ecophenomenologists like the above who, by and large, pick up the flesh ontology because they're interested in its intrinsic relationalism as a means of addressing environmental problems. When appropriated by other thinkers less interested in this aspect of its endless potential extensions, however, the cast of mind involved in thinking about the reversibility of the flesh as a site of justification for one's reflective conclusions appears to threaten the possibility of a suitably reflexive human-world praxis. My basic point is this: the flesh *itself* doesn't appear to be doing any positive or useful work over and above the chiasmic ecophenomenology offered in Merleau-Ponty's earlier works.

The foregoing should come with a few qualifications. I'm not claiming that ecophenomenologists do not employ arguments in support of both the flesh ontology and their respective appropriations of it. All the aforementioned ecophenomenologists advance quite sophisticated (although not unproblematic) cases that *their* position results from serious consideration of the sort of radically flattened, intrinsically relational ontology instantiated in the flesh.[24] But, even where arguments are offered, we're left with the same two persistent and interrelated issues. The first is that the flesh ontology tends to limit the possibility of our critical investigation of it.[25] Of course, one can point out controversial assumptions (e.g., does Madison realize that, as global capitalism ironically demonstrates, individual flourishing *needn't* require the flourishing of all unless one adopts a circular or trivial definition of "flourishing?"), or fallacious logical moves (although ecophenomenologists almost never claim that the flesh *entails,* rather than *implies*, their recommendations). In lieu of a yardstick, this is as far as one can go. But, to reiterate, these critical tools don't require the flesh; they're available within Merleau-Ponty's earlier model and so the flesh itself offers little extra merit for the project in hand.

In any case, situating one's capacity for reflection within the flesh's auto-affection does little to promote critical self-reflexivity about the assumptions which underpin those arguments. This problem isn't limited to extreme cases like Abram's. Resituating the human-nature relationship within the flesh appears to do little to make Jensen, for example, any *more* attentive to the fact that, by beginning her analysis at the species level, she appears to unquestioningly instigate a hierarchy of value which is set against the recognition of individual entities. Nor does the flesh appear to make Sally Fischer any *more* reflexive in her uncritical promotion of the ecologist Joel Kovel's contention that "there is no separation between things; there are only, so far as the most advanced science can tell us, plasmatic quantum fields; one single, endlessly unperturbed, endlessly becoming body" (Kovel, cited in Fischer 2007, 206). Fischer disrupts Cartesian mechanism's instrumentalist stance, but only by *desituating* or uncoupling the ontology licensed by quantum physics from any phenomenological opening or sociohistorical entanglements. In neither case does the flesh ontology facilitate any obvious improvement in terms of genuine openness to alterity or wonder.

As I articulate our overarching problem above, by entirely subsuming subject and object poles into the flesh's autoaffection or its continual becoming, the risk is that Merleau-Pontian ecophenomenologists become more (or, at least, no less) prone to distract their attention from acts of critical self-reflexivity. This is plausibly because, in short, that "self" is understood as only a secondary abstraction from the flesh's larger self-becoming, *just like everything else*. The reader might be forgiven for thinking that we've been here before. In chapters 2 and 4, I criticized ecofeminists and speculative realists for making similar errors. I criticized Warren, Merchant, and Plumwood for advancing similarly flattened ontologies under which nonhuman interlocutors are misleadingly identified as suspiciously similar intentional subjects. Object-oriented ontologists, I argued, make the self-same error by making all objects (humans included) the same kind of intentional entity. In each case, I argued that the real reason for their hostility to alterity is that, by universalizing the terms of their respective fundamental ontologies, they obscure the fact that any such ontology is always derived from a situated phenomenological opening which inevitably colors its norms. Since any subsequent ontology is then given fundamental status, the specificities of that sort of opening are glossed over as "general," yet may conceal a certain (potentially problematic) anthropomorphism or ontological anthropocentrism at root. They all risk failing to recognize that, even when our aims in doing so are noble, it matters *who* flattens the ontology and *under what terms*.

Of course, the ontological flattening characteristic of the flesh doesn't follow *exactly* the same pattern as progressive naturalism or object-oriented ontology. Those who employ the flesh ontology at least ordinarily acknowledge the necessity of approaching ontology from a particular type of phenomenological opening (unlike the former) *and* that other openings may be radically and irreducibly different from my own (effectively unlike both former and latter). Nevertheless, the flesh ontology *does* court two parallel errors. In some of its guises, it first risks making the more-than-human world intentional in more or less the same manner that we saw Bannon criticize Plumwood for doing in chapter 4. By situating the motor of intentional expression within the world itself (perhaps with specific things "it wants to say"), on the monistic reading, at least, Merleau-Ponty appears to asymmetrically *person*ify it in a nonrevisable manner, rather than affording the possibility of rethinking or addressing it on its own terms. Second, by giving the (albeit indirect) flesh ontology priority in subsequent argumentation, one may be driven to obfuscate or neglect the concrete phenomenological opening(s) under which the ontology was flattened. Thus, the intentions and "general" relations that one finds in the flesh will likely be intimately related to the sorts of onto-epistemic entanglements typical of the partial opening(s) under which the flesh was diagnosed or interrogated and yet, because of the apparent primacy of the flesh, one appears more vulnerable to not noticing. Conceived as such, the ontological shift at play in the mature Merleau-Ponty does little to encourage the critical self-reflexivity which is conducive to a radically reflective praxis.

6.5 CONCLUSIONS

In this chapter, we explored whether turning to the flesh might benefit the nonanthropocentric credentials of the ecophenomenological praxis I defend. Broadly speaking, we explored two potential benefits: first, that, in support of a less anthropomorphic or ontologically anthropocentric view-from-everywhere, the flesh might facilitate an analogous dialectic with nonhuman entities that an intersubjective hyperdialectic provides between human body-subjects; second, that, regardless of any normative guidance that the flesh might provide, the cast of mind encouraged by resituating the human-nature relationship within it might provide a helpful push away from ontological and/or ethical anthropocentrism in our reflections. I've argued, however, that, by understanding one's expressions to be constitutive of the flesh's autoaffection or self-becoming, one may become less likely to focus on our tacit contributions to those expressions. Without the possibility of some further, usable tools by which to curtail our tendencies toward onto-epistemological imperialism, therefore, the turn to the flesh may even prove a retrograde one in terms of our genuine openness to alterity because we may become *more* likely to subsume the more-than-human world under the terms of our partial purviews. Thus, especially given that the primary benefit of our ecophenomenological praxis is the sort of critical self-reflexivity that risks being diminished by a turn to the flesh, we should have grave reservations about taking up the flesh ontology as part of any such praxis.

"CONCLUSIONS"; OR, BECOMING A PLACE OF UNREST

The conclusion reached in the previous chapter might prove somewhat disappointing because, if no straightforward equivalent for an intersubjective hyperdialectic can be established with the more-than-human world, then there appears to be an inescapable ontological anthropocentrism inherent to the praxis I defend. Moreover, a critic might claim, if disrupting the inherent anthropocentrism (in the more troubling utilitarian, impositionistic, or chauvinistic senses) of our dominant onto-epistemological schemas is as important as I suggest in addressing the root causes of our crisis situation, then my praxis is not only bound for failure, its failure is bound up with the *seemingly* paradoxical task I set for it. What I want to show in this final chapter is that my hypothetical critic's worries can be allayed in such a manner as to limit the extent of any disappointment that the reader might legitimately feel. While my critic rightly claims that the praxis I defend is, in *some* sense, anthropocentric, I shall show by means of a substantial recap that it is in engaging with its ontological anthropocentrism head on that an ecophenomenological praxis shows its mettle in disrupting the habitual inattentiveness that is complicit in our colonial appropriation of the more-than-human world. To unpack this claim, I will focus here, as I have throughout, on how one's embodiment provides the means by which one both is, and is not, to "return to the things themselves."

Throughout this book, I've problematized attempts to respond to the environmental crisis that begin by construing it as a series of environmental "issues" (determined by natural scientists) to be "solved" by the right kinds of practical behaviors (the stipulative realm of environmental ethics). By ceding ontology and epistemology to natural science in this manner, I've argued, we leave the philosophical and political assumptions and commitments which underpin it largely unchecked. At least in the context of environmental philosophy, I've suggested, any subsequently plausible solutions to environmental "issues" are thus prone to inherit the normative violence which imbues the limited and utilitarian models that natural science offers of those phenomena. If, as I claim, this conceptual violence is intimately linked to the behavioral and attitudinal violence enacted in aggressive deforestation, excessive fuel consumption, overfishing, and so on, then any such "solutions" will be of limited merit in disrupting the anthropogenic violence which precipitates our crisis situation.

I've suggested that scientific naturalism's problems stem largely from its underpinning objectivistic assumptions and commitments, which limit its capacity to identify and disrupt the violence implicit in the models it offers of things. The root of these problems lies in naturalism's fundamental commitment to objective thought, which motivates the natural scientist to mistakenly reify a purified "objective" realm of objects or properties-in-themselves and, thus, conceals the irreducible contribution of the situated body-subject (i.e., the scientist) to the concrete physiognomy of its underlying phenomena. By effectively limiting ontological and epistemological enquiry to the modality of the in-itself, natural science risks naturalizing the terms of epistemic engagement specific to a restrictive and partial purview in a manner that doesn't foster critical self-reflexivity about its underlying assumptions and commitments.

What ecofeminist analyses add to the phenomenological critique of the metaphysical dogma manifested in scientific naturalism is a clear focus on the *political* undercurrent of its objectivistic commitments: an undercurrent set against recognition of nonhuman entities (and other Others) on anticolonial terms. For ecofeminists, the major problem with taking the objects typically revealed by natural science to exhaust ontological reality isn't just that the scientist mistakes himself to be fundamentally passive in receipt of the world's truths. The issue is rather that, in order to construe nonhuman entities *as* objects in the requisite, totalizing sense, the ways in which the requisite "passivity" is granted to him conceal an implicit hostility toward nonhuman alterity and a problematic propensity toward the sort of onto-epistemological violence of which Bacon is the exemplar. Val Plumwood, for instance, gives reason to believe that the inherent *dualism* of the terms under which allegedly plastic or inert material "objects" are cleaved from minded human "subjects" is partly to blame for the instrumentalist stance under which the more-than-human world is subsequently subsumed by those who, like Paul Crutzen, *reduce* it to a collection of "natural *resources*." Thus, ecofeminists and ecophenomenologists argue, successful attempts to address environmental "issues" cannot simply revolve around which normative policies and maxims to apply to the phenomena that scientists describe. They must also involve disruption of the problematic (meta)theoretical assumptions, commitments, and apparatuses which dictate the "environmental issues" that natural scientists bequeath us, and which otherwise constrain what plausible "solutions" might look like (in terms of more sustainable models of "resource consumption," and so on).

Ecofeminists like Plumwood are surely right to argue that a key strategy in disrupting the latent dualism manifested in scientific naturalism is to problematize the reductive and hierarchical subject/object binary upon which it relies. Despite Plumwood's calls to reconceive of ourselves as more substantially embodied than the dominant "truncated reversal" of Cartesian metaphysics otherwise licenses, however, a problematic silence persists among ecofeminists about the *concrete* role that one's embodiment plays in more fundamentally configuring the world that one inhabits. The problem with this oversight is that, by failing to engage sufficiently with the salience of one's specific situation *as* an embedded body-subject with a partial grip on the world, ecofeminists are often insufficiently radical in their revision of terms. Plumwood, for instance, makes two fundamental errors which unduly constrain her attempt to rethink the terms under which environmental discourses may take place. First, the terms of her progressive naturalism do not admit of the radically heterogeneous intentional realities which may be inhabited by differently embodied entities. While Plumwood is open to rethinking the subject/object dualism in the form of subject-subject relations, the terms of these new relationships are still nonaccidentally aligned with the narrow purview of the human investigator and, therefore, remain in tension with a radical rethinking of how to address nonhuman entities (like snakes) *on their own terms*. Second, partly because of her insufficiently radical rethinking of "subject" and "object" terms, Plumwood misunderstands the basic error of dualism to be a conceptual one. Plumwood, like other ecofeminists, is therefore led to underplay the true significance of understanding dualistic schematization to be a more-than-cognitive error and, thus, misconstrues how best to go about addressing the colonial habits responsible for our crisis situation.

I've argued that this is where significant gains can be made by reconfiguring ecofeminist (and new materialist) insights within a Merleau-Pontian ecophenomenological framework. For the Merleau-Pontian ecophenomenologist, one's phenomenological opening is always already subject to the habitual sediment of one's situation as well as the operative demands of one's specific mode of embodiment. These admissions afford an important tonic to ecofeminism's shortcomings because they allow the Merleau-Pontian ecophenomenologist to foreground the *intentional* contribution of one's embodiment to the basic meanings that the world presents. As a result of a keener focus on the salience of one's embodiment, the radically reflective Merleau-Pontian ecophenomenologist may be less prone to impose onto-epistemological violence onto the more-than-human world for at least two related reasons. First, because under any such ecophenomenology a body-subject is, as a matter of principle, attuned to the possibility of dialectical denial by the world. This disrupts the sort of ontological incorporation or habitual instrumentalism which continues to permeate (even a "progressive") naturalism. Second, because Merleau-Pontian ecophenomenologists' intractable Gestalt-situatedness makes the error involved in objectivistic or dualistic habits a more fundamental one of *praxis*. Given one's embodied situation, this does mean that a complete transcendental reduction is rendered impossible as a means of identifying and rectifying the problematic metatheoretical terms of debate. I've argued, however, that the tension Merleau-Ponty exposes between one's immanence and transcendence not only motivates a stance of humility that better facilitates openness to more-than-human alterity but also allows the Merleau-Pontian to focus on the level on which to attend to the problem of dualistic schematization: the level of perceptual habits themselves.

By recognizing that the *meanings* coexpressed between body-subject and world are ontologically basic, and that the meanings available for expression are nonaccidentally related to the specific kinds of situated bodies that we are, the Merleau-Pontian ecophenomenologist is able to embrace a radically anticolonial onto-epistemological norm which may reduce the violence done to more-than-human alterity without deferring to problematically objectivistic notions of how stipulative things-*in-themselves* really are.[1] This comes in the form of a view-from-everywhere which may encompass the intentional meanings differentially expressed by divergent more-than-human entities. Nevertheless, since the concept of expressive truth central to this norm makes hoping for perceptual or conceptual coincidence with the "essence" of a given more-than-human thing impossible, the above might be taken to exemplify how, for the Merleau-Pontian, one's perceptual embodiment actually *prevents* a return to the more-than-human things themselves. This is the central claim of the charge of correlationism in both general and sexually specific guises.

If we must abandon the modernist schema of relations between passive subject and object-in-itself as ecofeminists and phenomenologists rightly suggest (for different, though compatible reasons), then (environmental) philosophy appears to be reduced to investigation of the epistemic relations between a given body-subject and the (properties of) objects which are its correlate. Given that the "correlational" properties thereby revealed are non-accidentally related to the sort of body that one is—or, given that, for any specific body-subject, "the its outnumber the mes" (Bennett 2010, 112), the bodies that one is—as a situated investigator, one risks being divorced from a world of genuine Others. According to the speculative realists, the radical alterity made plausible by a phenomenological appreciation of the intentional salience of one's embodiment is also simultaneously foreclosed by the impossibility of transcending the narrow terms of body-world relationships available only to perceivers *like me.* If our dualistic and objectivistic tendencies delineate how we habitually carve up the world, then ecophenomenologists may also face a narrower problem whereby *androcentric* sediment colors the allegedly primordial epistemic norms of even "perceivers like me" in such a way as to perpetuate, rather than challenge, the violence at the core of our enquiry. In chapter 5, we saw how this thought played out via the objectifying stance naturalized by Husserl's transcendental reduction.

Since the various ways in which one might understand the meaningful scope of environmental theory from within the correlationist circle (e.g., via William Cronon's or Steven Vogel's constructivism) concede a problematic hostility to nonhuman alterity, the correlationist problematic is a serious one. Nevertheless, I've also argued that attempts to speculatively transcend the bounds of the correlation are hostile to the radical alterity that more-than-entities must retain under any appropriate reconfiguration of ontological and epistemological terms. Speculative realist attempts to resolve the epistemic asymmetry of subject/object relationships either implausibly "nullify" the intentional contribution of the investigative body-subject or else—not unlike some ecofeminists and, perhaps, even the later Merleau-Ponty—*flatten* ontology such that the way things play out for a given human "subject" provides the "general" terms for epistemic engagement between all entities. However, insofar as they permit epistemic access to more-than-human entities at all, speculative realist approaches are prone to extend to only the (reflectively distorted) terms of our own worldly experiences and, thus, conceal the contribution of the specific opening from which any such terms are gleaned. We saw how this played out in Quentin Meillassoux's neo-Cartesianism and

Graham Harman's oscillation between a homogenizing anthropomorphism and a problematic absolute alterity. Speculative realists (like most scientific naturalists, ecofeminists, and the later Merleau-Ponty) risk thereby doing violence to more-than-human alterity by failing to take the intentional importance of one's specific mode(s) of situated embodiment sufficiently seriously.

At the risk of laboring an important point, the above is intended to expose two apparent paradoxes central to Merleau-Pontian ecophenomenology.[2] One is the insoluble paradox of immanence and transcendence which we explored in chapters 1 and 5. Since all ontological and epistemological investigations take place in the intentional horizon opened up by one's embodied engagement with(in) the world, one's phenomenological embodiment provides the basic means by which one *must* return to the things themselves if one is to evade schematic terms which are reflectively worked-over in a misleading and problematically anthropocentric or androcentric manner. Under this paradox, however, the phenomenologist cannot "get back" to the meanings primordially permitted by the more-than-human world, because one is always concretely embedded within a sociohistorical lifeworld which rules out the sort of transcendental reduction that might otherwise bring all of one's intentional commitments to one's attention. Thus, even a radically reflective praxis will never be free from intentional sediment in such a manner as to allow unmediated or transcendental acquaintance with the more-than-human things themselves.

Our second apparent paradox has no catchy name, but it relates to the relationship between specific embodiment and radical alterity. For the appropriately engaged Merleau-Pontian ecophenomenologist, one's specific mode of embodiment cannot be a "general" one that can be extended unproblematically to Others. It is this focus on the intentional salience of one's material embodiment which brings to our attention the possibility of radically heterogeneous, meaningful worlds potentially inhabited by more-than-human Others. However, given that one's immanence is intractable, one cannot transcend one's own embodied situation to exhaustively engage with the divergent worlds of more-than-human Others on the very radical terms that attention to embodiment makes possible. This is the problem involved in any attempt to naturalize the Gestalt structure from "outside," a problem which Merleau-Ponty diagnosed in his first book. According to our second apparent paradox, the radical nonanthropocentrism of onto-epistemological engagement manifested in any view-from-everywhere appears to be compromised because it will always be somewhat inflected by the operative or habitual norms of the asymmetrical phenomenological opening(s) from which it is considered.

My reason for laboring the point about the apparent paradoxes of ecophenomenology is that I suspect that in them lies the root of my hypothetical reader's disappointment. On the face of things, both make the same basic point: it is through attention to embodiment that Merleau-Pontian ecophenomenologists make radical engagement with more-than-human alterity theoretically possible, yet unobtainable in practice. As some recent ecophenomenologists have suggested, an ontology of the flesh promises to mitigate our concerns by situating one's intentional expressions within the world's own becoming. However, as I argued in chapter 6, any reversibility that the flesh could legitimately afford facilitates no real improvement on this front. In fact, by taking up the flesh ontology, one perhaps becomes more likely to repeat the error made by speculative realists and ecofeminists under which, by flattening ontology according to the terms of a "general" or "anonymous" perspective, body-subjects become less critically self-reflexive about their acts of violence against more-than-human alterity. In short: through the sort of approach that would, on the face of things, provide a tonic for the background ontological anthropocentrism inherent to ecophenomenology, one risks attempting to make the Other's unfamiliarity too familiar and thus limiting its potential to identify and disrupt the colonial violence characteristic of our crisis situation. Ecophenomenology, it would seem, is then unavoidably anthropocentric in at least a reasonably strong onto-epistemological sense.

While I don't wish to deny the terms of our two apparent paradoxes as I've set them up, I do want to argue, as I have throughout this book, that the focus of the preceding debate is too narrow. What we've focused on is the task of ecophenomenological critique: the content of the second strand of radical reflection (reflection in the world). Another key benefit of the work done by ecophenomenologists, ecofeminists, and new materialists in problematizing the mind/body, self/other, and subject/object dualisms is to reveal a fundamental connaturality between human and more-than-human parties. The ontological continuity facilitated by the body-subject's carnal embodiment undermines the correlationist charge against ecophenomenology because, even as a situated body-subject, one simply isn't confined to the "subjective" epistemic prison implicit in its more troubling manifestations. Although one cannot be permitted unmediated, exhaustive, "access" to the more-than-human world's own truths in the manner of a Baconian observer, as a Merleau-Pontian body-subject, one is to some degree implicated in the world investigated and the heterogeneous intentional perspectives of more-than-human Others therein via common nondiscursive grounds of experience. Ascertaining the "brute presence" of the world in this sense is the focus of our other strand of radical reflection (the world in reflection). Thus, while it is true to say that ecophenomenology is unavoidably anthropocentric in the sense that the world's normative meanings are inescapably revealed to and imbued by the onto-epistemological norms of a human phenomenologist, as Merleau-Ponty, Plumwood, and Jane Bennett all testify, one is neither entirely nor exclusively human in the requisite sense for this admission to result in the species solipsistic conclusions that speculative realists claim.

This leads me to why I think that the failure of a turn to the flesh shouldn't prove disappointing for the ecophenomenologist who wishes to do justice to more-than-human alterity through a radically reflective praxis. I've argued that, while a rigorous ecophenomenology may reveal sufficient nondiscursive intertwinings between more-than-human perspectives to afford varying degrees of engagement with them, taking seriously the limitations of one's embodied grip suggests that, as I noted above, the Other really is *unobtainable*. But this shouldn't be a cause for regret since, as Plumwood argues, the normative impetus to exhaustively possess other entities, even in this epistemological sense, is shot through with the very dualism we seek to disrupt. The self-same normative political impetus is, after all, partly to blame for the conceptual violence done via our mistaken adherence to the objectivistic lens of scientific naturalism and the behavioral violence that purview licenses. What Plumwood argues, instead, is that a more positive politics requires us to understand all theory to be partial and situated, rather than transhistorical and acosmic. The real contribution Merleau-Pontian ecophenomenology brings to this picture is simply—in virtue of its keen focus on the fundamental ambiguity of embodiment—to afford a more radical alterity at the same time as affording a legitimate narrowing of the distance in partiality of perspectives between self and Other. If the main value of Plumwood's metatheoretical project lies in her attempt to take seriously both continuity and difference, then hers is a project that the Merleau-Pontian ecophenomenologist—through the figure of the chiasm that is so important to Merleau-Ponty's work—may take up more effectively.

To clarify: permit me to pick up on the spatial metaphors that have punctuated this book. By *cleaving* subject and object poles, naturalists ordinarily fail to engage important continuities or isomorphisms between human and more-than-human parties, and so do undue onto-epistemological violence to the underside in a manner that we see manifested in anthropocentric instrumentalism at the level of normative guidance (e.g., via Crutzen's and Manuel Arias-Maldonado's imperatives to rectify our crisis situation by *mastering* the more-than-human world). I've argued that this holds true, even under the truncated reversal ordinarily manifested in the extensional realist stipulations of natural science (think back to our earthworms and plant intentionality wars). Those who take up otherwise *flattened* ontologies of the ilk discussed in this book (i.e., some speculative realists, ecofeminists, and later Merleau-Pontians), on the other hand, often fail to engage sufficiently with important differences or asymmetries between those parties and so are apt to carry out related acts of violence via the "general" extension of the terms of specific sorts of embodied experience. The radically reflective Merleau-Pontian ecophenomenologist, however, offers something like a topology of *uneven ground* under which, although there are passages between features (some easier to navigate than others), there are also aspects of the landscape which are, either in part or in whole, in-principle inaccessible to me because of the kind of ambulatory feature that I am (think about Merleau-Ponty's canines or Bennett's plastic). But these features nevertheless remain part of the landscape and demand to be considered as such in any debate about how to conceive, approach, or value that landscape, especially when the landscape, in whole or in part, is under threat.

Yet even on this ecophenomenological landscape, all is not always as it seems (I will drop the metaphor here to avoid stretching it beyond all credibility). As I argued in chapters 1, 3, 5, and 6, one cannot think that connaturality affords easy acquaintance either with the world's meanings, or with those shared with more-than-human Others, without violating the requirement for reflexivity about one's own sedimented and operative intentional commitments. This is why, in order to do justice to more-than-human alterity, the body-subject's relationship with the rest of the world must remain unrestingly hyperdialectical as a matter of praxis. Of course, by ruling out an appeal to the flesh, this hyperdialectic cannot be a straightforwardly reciprocal dialogue with the more-than-human world. But, if my reasoning thus far holds, this "failure" may serve to better establish the sort of critical self-reflexivity that is most positive about the ecophenomenological project. Although, as a body-subject, one is connatural with the rest of the world in the requisite manner to afford communication with it, one cannot transpose the terms of an intersubjective dialectic onto one's relationship with the *more-than*-human world without misrepresenting the partial, chiasmic opening one has onto it in virtue of the kind of situated body (or bodies) that one is. To do so would be to attempt the sort of "shortcut" that Sullivan rightly associates with the very colonialism we're employing ecophenomenological tools to disrupt. Contextualized as such, Merleau-Ponty's later ontological turn comes across largely as an attempt to overcome the asymmetrical terms of the human-world chiasm which, almost paradoxically, may represent a retrograde move in the attempt to appreciate more-than-human entities in their alterity through a view-from-everywhere.

Nevertheless, it would be a serious blow to our project if, given the residual epistemic asymmetry of the chiasm between body-subject and world, one was left without any critical tools with which to unearth nondiscursive grounds of experience and expose discursive sediment. Such an admission would be fatal for the promise of our new onto-epistemological norm, given that the view-from-everywhere would be reduced to a problematic sort of de facto correlationism, in which our intractable ontological anthropocentrism is intimately related to a more straightforwardly pernicious ethical anthropocentrism. But I've argued that this claim isn't justified either. Throughout this book, I've suggested multiple hyperdialectical means (e.g., radical reflection, habitual hesitation, [targeted] intersubjective dialogue, inter- and intradisciplinary dialectical fecundities) by which body-subjects might address their colonial imposition of (meta)theoretical assumptions and commitments onto divergent more-than-human entities. Of course, given the asymmetry and partiality of one's opening onto the world, the focus of these tools is primarily critical, but this admission shouldn't be taken to undermine the nonanthropocentrism of my praxis in any problematic (e.g., ethical or straightforwardly ontologically impositionistic) sense. Rather, the habitus of critique which is focal to the Merleau-Pontian ecophenomenological praxis I endorse is testament to the seriousness with which it acknowledges the inherent tensions within the fact that any metatheoretical groundwork must be carried out by a situated and embodied entity that cannot entirely transcend the terms of the world that they inhabit.

If environmental philosophy is to be critically self-reflexive in the requisite manner to be open to more-than-human alterity, then this appears to be a tension with which we must reckon. But we should remember that there are real benefits to our ongoing praxis of taking it to heart. Doing so, I've argued, is also more likely to afford a habitual onto-epistemological humility—or, as Merleau-Ponty puts it, doing so allows one to become a "place of unrest"—which sets the body-subject in good stead to attend to more-than-human alterity in a suitably radical fashion. Conceived as such, the grounds of Merleau-Pontian ecophenomenologists' peculiar kind of nonanthropocentrism lie in the fact that they "limp" as a matter of praxis because their commitment to radical reflection hinders their commitment to unreflective practical action. Nonetheless, since addressing the violence implicit in the ways that those "issues" are ordinarily set up is an important part of disrupting the behaviors responsible for our crisis situation in the first place, "the limping of philosophy is its virtue" (*IPOP*, 61).

NOTES

INTRODUCTION

1. I retain the language of environmental *crisis* to emphasize its immanent, all-encompassing, and troubling nature. While a crisis so understood may require urgent *attention*, its demand for engagement (rather than relatively uncritical practical action) needn't inhibit a critical approach to the ways it is typically configured.

2. One might have reservations about Evernden's reduction of the "submerged mass" to a domain of "assumptions," given that some of its constitutive commitments may be more-than-cognitive. However, this observation remains consistent with Evernden's major point that the environmental crisis is linked to the *assumption* of an objectivistic, utilitarian, and/or scientistic purview.

3. I'm happy to court parallels with Slavoj Žižek's views on violence. Žižek (2008, 2) argues that there exists a "symbolic" and "systemic" violence which underpins visibly violent behaviors like the 9/11 terrorist attacks, but remains "invisible since it sustains the very zero-level standard against which we perceive something as subjectively violent" (i.e., violence enacted by a discernible subject). Žižek also cautions against the urgent injunction to "rectify" violent *actions* because, in doing so, we leave the background violence untouched.

4. See Callicott (1989) or Fox (1990).

5. See James (2007, 502).

6. Unlike Toadvine (2009) and Erazim Kohák (1984), I don't eschew the term "environment" because I want to retain resonances with Uexküll's contentions about our contributions to the world's physiognomy.

7. Putnam contends that the "*externalist*" (1981, 49) perspective is regretfully prevalent in analytic philosophy. Although Putnam's contention doesn't sit easily with my story about post-Kantian philosophy, it makes no difference to my wider argument. Putnam's point is that contemporary philosophy is problematic where it retains the dogmatic metaphysician's naive correspondence theory of truth.

8. Vogel here parallels Manuel Arias-Maldonado (see chapter 3). Interestingly, given the Uexküllian context of the present concern, Vogel contends that "this character of our artifacts—their reality and difference from our ideas and intentions—is not 'beyond' us any more than spiderwebs are beyond the spiders that spin and inhabit them. The moment of 'independence' in our artifacts, or in the spiders' webs, is simply the moment of realness, and in that sense is analytically implied by the very notion of making something, or for that matter of acting in the (real) world" (2015, 168).

9. Toadvine (2009) makes related claims about the requirement to navigate the demands of metaphysical realism and idealism in any "philosophy of nature."

10. This basic distinction between ontological and ethical anthropocentrism is Tim Hayward's (1997). Hayward contends that either can be benign or objectionable depending on how that anthropocentrism is manifested and that, while ethical anthropocentrism isn't strictly entailed by ontological anthropocentrism, the two are intimately related.

11. Although I cannot expand on this thought here, Merleau-Pontian ecophenomenology enjoys advantages over other (e.g., Heideggerian or Husserlian) (eco)phenomenological traditions by focusing on the irreducible contribution of the *specifics* of one's embodiment to the world's physiognomy. Heideggerian "Daseinanalysis," for instance, is frequently criticized for underplaying the *ontological* significance of bodily specificities, which Heidegger generally sees as "contingent and accidental configurations without transcendental relevance" (Heinämaa 2018, 544; Sartre 1969, 382–83). Insofar as he retains a transcendental phenomenology, Husserl displays a related hostility to the intentional heterogeneity enjoyed by differently embodied entities, a failing his supporters are apt to inherit (see chapter 5). The later Husserl only began to address this shortcoming by embracing a phenomenological project remarkably similar to Merleau-Ponty's (see *PP*, vii–xiv).

12. Hayward (1997, 53–54) equates this kind of ontological-cum-ethical anthropocentrism with human chauvinism.

13. By "diacritical relationships," I mean the various interrelationships between (otherwise indeterminate) aspects of one's perceptual field which codetermine the qualities of the intentional objects that show up perceptually.

14. I associate the former with the "externalist" sense of transcendence which Putnam links to a God's Eye view. The latter implies a Kantian or Husserlian sense of transcendence, under which, roughly speaking, one purifies conscious experience of all "inessential" conceptual components (e.g., by bracketing the contribution of one's political allegiances, if not those of space and time).

15. A chiasmic relationship is one of partial encroachment. I explore the concept further in chapter 1.

CHAPTER 1: PERCEPTION AND UNREST

1. I take "naturalism" to be a metaphysical position which eschews "supernatural" properties and limits ontological and epistemological inquiry to the modality of the in-itself, ordinarily, as described by (an ideal) natural science. "Objective thought" is more fundamentally a theory of perception which takes its basic constituents to be determinate objects or properties. These terms are intimately related. Naturalistic metaphysics appears to derive, in part, from a misplaced adherence to objective thought. "Natural science" refers to those disciplines which aim to provide means of exhaustively describing or explaining "nature-in-itself" (variously conceived). Although natural science needn't necessarily make substantive ontological pronouncements, natural science usually serves to describe the basic metaphysical structures upon which it already implicitly relies. Given their multidirectional transfer of influence, it's difficult (and perhaps disingenuous) to consistently separate natural science from naturalism.

2. Some (e.g., enactivist) approaches don't formally subscribe to either paradigm. I touch upon the objectivistic residue in these accounts in chapter 6.

3. According to "reflex arc" accounts, motor behaviors are wholly explicable through mechanical relationships between external stimuli and "objective" physiological phenomena in a neural pathway. Although reflexes may be conditioned, any such conditioning remains a mechanical relationship between the allegedly determinate entities involved. Reflex arc theorists (e.g., Pavlov 1927) typically eschew behavioral explanations which retain essential reference to intentional or existential significances.

4. Given its naturalistic limitations, science cannot *obviously* account for the experiential subjectivity upon which its possibility rests. As the sway toward reductive forms of naturalism in metaphysics and philosophy of mind testifies, a central goal of science has been to explain consciousness *away* because of its incompatibility with scientific apparatuses.

5. Not all do. Niels Bohr, for instance, refused to ontologize the findings of his quantum physics. For Bohr, "objectivity" refers only to the possibility for unambiguous communication between human scientists who share certain metaphysical and cultural prejudices. However, as Chris Calvert-Minor (2013, 127–28) observes, Bohr's rejection of objective thought comes at the cost of being unable to provide a nonanthropocentric (meta)physics.

6. Although some (e.g., Ihde 1993) dispute whether technological Enframing is specifically modern, the literature is in little doubt that it is facilitated largely by modernistic sediment.

7. Like Plumwood, I use the term "humanoid" to problematize any straightforward human/nonhuman binary operating along the fault line of consciousness. While some nonhuman entities (e.g., "higher" primates) might possess thematizing consciousnesses, there is no good reason to think that consciousness is limited to these humanoid determinations.

8. As Helen Steward (2016) emphasizes, however, our best theories of physics (e.g., nonequilibrium thermodynamics) profess no official commitment to determinism (nor indeterminism). Commitment to determinism is probably metaphysical in a way that will never be shown to be true by science itself.

9. See Varela, Thompson, and Rosch (1991), Clark (2008), or Thompson and Stapleton (2009).

10. An optic chiasm is where optic nerves partially cross over in one's brain. Although the nerves aren't themselves photosensitive, their crossing-over is essential for binocular vision. Just as monocular vision is abstracted from binocular vision, Merleau-Ponty thinks "subject" and "object" are reflective abstractions from their ordinary coconstitutive expressive contexts. For Merleau-Ponty, *partial* encroachment between chiasmic poles, and not their juxtaposition or assimilation, is ontologically basic (*PP*, 93; *VI*, 7–9, 123, 148).

11. Merleau-Ponty claims that the animal milieu is characterized by "the monotonous *a prioris* of need and instinct" and the human milieu by "use objects" and "cultural objects" (*SB*, 162). However, this rigid and potentially chauvinistic cleavage is inessential to his broader argument. Merleau-Ponty himself cautions against restricting *types* of perception to types of organism (*SB*, 104).

12. Smith's translation conflates "body image" and "body schema": "'body image' refers to one's conscious reflection on how they see their own bodies and the type that it is for them. Our 'body schema' on the other hand is tacit yet unreflective knowledge

about our body, its movements, its generation of space—as being-in-the-world" (Mahendran 2007, 195).

13. "Connaturality" needn't make meanings univocal or universal. As a matter of phenomenal primacy, one cannot *absorb* different meanings available to Others into one's reflective consciousness (*PP*, 42, 62). This realization is essential to his point about the canine sexual pantomime.

14. A hyperdialectic denies the possibility of neat synthesis because "every *thesis* is an idealization" (*VI*, 94–95). "Bad" dialectics like Hegel's err in understanding dialectical meanings apart from their partial, situated, and concrete configurations.

CHAPTER 2: ECOFEMINISM AND ECOPHENOMENOLOGY

1. Talia Welsh (2008, 47–51) notes that, throughout the *Sorbonne Lectures,* Merleau-Ponty himself contends that the "abnormal" categorization of woman juxtaposed against the masculine "universal" is the result of sociohistorical sediment, which education perpetuates.

2. Even Plumwood (2006) is somewhat wary of confining dualistic thinking to the "West." Given the contemporary paradigm of aggressive cultural imperialism, "Western culture" (and, therefore, dualism) extends further than Plumwood consistently suggests. The "globalization of western agricultural practices" and prevalence of (nominally) free-market capitalism testifies to this (Glazebrook 2002, 18). Moreover, it is dangerous to romanticize "non-Western" cultures as being beyond the grip of a "Western" logic. Amartya Sen, for instance, argues that "the Western conceptualization of India as mystical and 'other' [actually] serves the purposes of the fundamentalist Bharatiya Janata Party who are busy rehashing history to serve the ends of their own political power" (Sen, quoted in Nussbaum 2005, 321f).

3. Plumwood's point isn't that specifically "male" conceptions enjoy privilege over "female" ones. It is rather "the complex cultural identity of the master formed in the context of class, race, species and gender domination, which is at issue" (1993, 5). Plumwood's target is a problematic *logic,* not one of its many manifestations. There is no essentiality about the alleged hierarchy or temporal priority of one kind of oppression over another (Plumwood 1993, 30–31). Thus, exposing dualistic logic may help us tackle the denigration of other Others (e.g., so-called nonheterosexual or nonwhite persons) in a manner consistent with (eco-)queer liberation strategies (Gaard 2015, 27).

4. This terminology is adopted from Marilyn Frye, who claims that "phallocentric reality" reveals itself in the "ontological conversion" characteristic of "the event of becoming a lesbian" (1983, 167, 171). Plumwood avoids the essentialism that Frye courts.

5. See Singer (1979) and Regan (1983).

6. Although Korsgaard (2005) extends duties to nonhuman animals, their value is entirely conditional on *human* legislative wills.

7. For Carolyn Merchant (1980, 233), capitalism also hyperseparates production (masculine) from reproduction (feminine), a division perpetuated via "gendered" employment. Plumwood (2002, 24) argues that the capitalist market is "freed" only through its hyperbolic disembodiment from ecological and social dependencies and is

thus facilitated by dualistic logic. The economic legacies of slavery (e.g., enforced reliance upon monocultures), for instance, pressure poorer nations to weaken expensive environmental constraints in order to participate in a global marketplace, even when the results are disastrous for the poorest. Similar issues pervade international emissions "trading." Vandana Shiva (1991) argues that this shouldn't surprise us, given that, through *their* dualistic identification with "mere nature," "non-Western" laborers usually fare worse than women in "Western" nations.

8. Difference feminists aren't necessarily committed to uncritical reversal strategies, since they usually problematize the mind/nature cleavage. This is true of Luce Irigaray, who nevertheless faces tricky questions about her essentialism (see Stone 2003). Identifying ecofeminism with *early* ecofeminism is one source of the mistaken charge that *all* ecofeminism is biologically or descriptively essentialist (e.g., Morton 2010). This charge also arises because ecofeminists must allegedly begin from a concrete male/female binary to establish their critique. However, ecofeminists needn't claim that dualistic terms track reality-in-itself, so it is misdirected. The difficult question about how then to identify subordinated groups without reinstating essentialism can be mitigated via two considerations. First, although the terms "male" and "female" may be politically useful, there needn't be a *clear* line of demarcation between them, nor need there be only two sexes. Second, dualistic master/slave distinctions may revolve around any number of identifiable features which, Alison Stone (2008, 155) notes, do happen to cluster. Self-identifying women are often breasted and/or share hormonal similarities, for example, but, dualism aside, these features needn't be essential or pivotal. However, since, under dualism, the markings of people's bodies are nonaccidentally related to their oppression, Stone argues, women will probably share significant overlaps in the genealogy of situation in which they take up their "womanhood." How women live their bodies within that situation may, therefore, afford sufficient family resemblances for an intelligible political group.

9. Although we can point to pivotal moments in genealogies of oppression, I resist the temptation to identify the *birth* of dualism. As Merleau-Ponty observes in the *Phenomenology*, the project of "transcending history" to exhaustively determine a movement's *causal* origins is mistaken. The pressing issue is how to identify and challenge dualism from *within*.

10. Barry Dainton (2008, 5) observes that we perpetuate such intuitions by telling children stories about princes and frogs "switching bodies" before they can scrutinize the conceptually laden language they inherit.

11. E.g., David Lewis (1966).

12. Exceptions include contemporary panpsychists (e.g., Galen Strawson [2006]), who broadly accept that physicalism exhausts reality, yet aim to retain the ontological irreducibility of consciousness. I address problems with panpsychism in chapter 3.

13. I rely here upon Karen Barad's account of discourse: discourse "is not what is said; it is that which constrains and enables that which can be said. [Discursive practices] are actually historically situated social conditions" (Barad 2008, 137).

14. See Heather Douglas (2016) for an excellent overview.

15. Some ecofeminists *do* reject metaphysical naturalism. Trish Glazebrook and Gail Stenstad (2011), for instance, are Heideggerians. For reasons previously outlined, however, I think a Merleau-Pontian approach is more promising for our task.

16. Alternatively, if Plumwood eschews representationalism, her naturalism might become consistent with Barad's (2006) process approach where the progressive

scientist's theoretical contentions aim to correspond with the whole experimental arrangement (Barad's "phenomena") rather than individuated things-in-themselves. Barad's naturalism takes the world-in-itself to just *be* its differential becoming and so promises greater openness to radically heterogeneous phenomena. As I argue (Forthcoming a), however, Barad thereby makes genuinely nonhuman phenomena *themselves* effectively inaccessible, or at least radically inconsistent with Plumwood's key claim about the "weakness" of the anthropomorphism involved in cross-species "representations."

17. I explore feminist reservations about phenomenology in chapter 5.

CHAPTER 3: SEEING BETTER

1. In the *Nature* lectures, for instance, Merleau-Ponty began to speak of the world's *own* intentional "*logos*" that he would later (mistakenly, in my view) characterize as the self-expression of "the flesh." Merleau-Ponty describes the developmental "unfurling" of axolotls, for instance, as being "like a pure wake that is related to no boat" (*N*, 176). Passages like these direct Bryan E. Bannon (2014) to contend that Merleau-Ponty later took up a Whiteheadian process metaphysics. For reasons explored in chapter 6, I'm unsure that Bannon's interpretation is consistent with the radicalization of phenomenology that motivates Merleau-Ponty's ontological turn. Nevertheless, the link is useful to support my concerns in chapter 6, given Plumwood's (2002, 128–31) contention that process metaphysicians perpetuate anthropocentric hierarchism by understanding humanoid properties like atomic consciousness to be universal norms to which nonhuman entities may be admitted by degree.

2. The two-page "Alpi Letter" (published as "Plant Neurobiology: No Brain, No Gain?" in *Trends in Plant Science* [Alpi et al. 2007]) was written by Taiz, Amedeo Alpi, and thirty-four other prominent plant scientists who objected to what they saw as a lack of intellectual and scientific rigor in the then-emerging field of plant neurobiology.

3. As Steven Shaviro notes, panpsychists usually focus on demonstrating that consciousness *couldn't* emerge because "all facts are determined by facts about the smallest things" (Coleman 2006, quoted in Shaviro 2014, 100). However, I've argued that "smallist" ontological hierarchies are androcentric and phenomenologically suspect. Furthermore, Bennett provides reason to think that sentience *could* emerge from nonsentient material, even from within a broadly naturalist purview. Since panpsychism doesn't appear to be a prerequisite for the existence of consciousness, our remaining question is whether it is palatable. I argue not.

4. Extending our charitable reading, Plumwood's "subject-subject" dialogue might just indicate an attempt to address the world on its own (sometimes nonintentional) terms. Like Merleau-Ponty, Bennett (2010, 108) thinks that doing so requires engagement with a *practical* dialectic, under which one is hyperattentive to the surprising results of nonhuman agency. This requirement chimes with Plumwood's promotion of "awe" in the face of the world yet needn't require all nonhuman parties to be "subjects."

5. Some panpsychists and object-oriented ontologists like Timothy Morton (2008, 77) notwithstanding.

6. "Assemblages" are bodies composed of relational multiplicities of fluid and impermanent actants (Deleuze and Guattari 2004).

7. Adorno thinks that naturalistic "identity thinking" (i.e., where one aims for conceptual adequation with the essences of "things-in-themselves") licenses the subjugation of "nature" "external" and "internal" to oneself. Like Merleau-Ponty, Adorno thinks identity thinking "compulsive" and prescribes hyperdialectical means of disrupting it (see Cook 2011, 67).

8. This might be unfair to Plumwood. She acknowledges relationality via her account of the intersubjective institution of dualism and how our material relationality requires us to widen the scope of prudential concern. Nevertheless, Bennett provides important details and a welcome shift in focus.

9. Alaimo (2007, 237) diagnoses a related aversion to "nature"—a so-called "accessory to essentialism"—common to feminism more broadly. The materiality of the body is typically recognized; she contends, however, that it is usually passive or plastic and, thus, amenable to construction, which, in post-Butlerian discourse, becomes the "true" determinant of identity. As Alaimo also argues, however, dualistically cleaving (nonhuman) nature from (human) culture isn't desirable either. Indeed, given its inattention to bodily resistance, alterity, and agency, "social constructionism can itself be thought of as a form of (discursive) essentialism" (Twine 2001, 37). I elaborate on this issue in chapter 5.

10. It would, however, be problematically anthropomorphic to locate *reflective* sediment like that of one's economic history within tree intentionality, for instance. Nonetheless, it would be too quick to entirely deny the impact of sedimentation, given the apparent import of kin recognition and intraforest communication to subsequent tree behaviors.

11. Alia Al-Saji identifies the following contributory categories: "diacritical, social, historical and material," but holds that, given our Gestalt-situatedness, "it is impossible to separate them" (2009, 377–80). My references to "diacritics" are shorthand for all the above.

12. Over the next two chapters I develop an account of nondiscursive grounds of experience (rather than full-blown nondiscursive experiences) to do justice to this issue.

13. For ticks, something similar might be said of butyric acid's "odor."

14. The occasional "shattering" of reality's normative content cannot, however, license a broader skepticism. The paradox of our "primordial faith" is that "there is the absolute certainty of the world in general, but not of any one thing in particular"; one's grasp of reality can be sacrificed only where a "truer" one may replace it (*PP*, 297, 360, 409). Likewise, "when I say that I see the ash tray over there, I suppose as completed a whole perceptual future. Similarly, when I say that I know and like someone, I aim, beyond his qualities, at an inexhaustible ground which may one day shatter the image I have formed of him. This is the price for there being things and 'other people' for us, not as the result of some illusion, but as the result of the violent act which is perception itself" (*PP*, 361).

15. As Kelly notes, however, while this is the inevitable outcome of Merleau-Ponty's analysis, he never commits to it unambiguously.

16. The rainforest may, of course, be different or multiple entities from some perspectives and no entity at all from others.

17. See Lawrence Brown (2015). Note also parallels with the emergence of recent zoonotic viruses.

18. I leave aside whether Merleau-Ponty's restriction to "*present* scrutinies" is consistent with what he says about temporal perspective (e.g., *PP*, 69). Ecophenomenologically speaking, the likely perspectives of future people should also probably inform how we presently encounter rainforests, for instance.

19. "Nature" doesn't here connote "untended wilderness" in the problematic sense implied by Crutzen and Arias-Maldonado. It refers to those things which haven't *obviously* been constructed with a functional capacity in mind. While "nature" and the "nonnatural" are both intersubjectively characterized, we should accept James's distinction as it relates to the difference between an Everglade swamp and a mobile telephone.

20. James (2009, 114–15) argues that while hesitation may hinder unflinching practical activism, such activism may only protect the *objects* of our concern: the things which are disappearing (e.g., rainforests, axolotls, and mountain gorillas). There is, however, a subtler loss quietly suffered whereby we subsume the more-than-human world into our objectivistic and utilitarian schemas, which hesitation may help to stem. Like Evernden, James thinks we cannot truly prevent the former loss without attending to the latter.

21. Morris makes a similar point in observing that, in a lizard's selection of a (hot, but not cold) stone on which to sunbathe, "an ontologically peculiar *but not* is key to the appearing of meaningful difference" (2018, 41).

22. Marginal awareness isn't fully thematized and so yields no straightforward intentional *object*. Retaining marginal awareness of inassimilable meanings and commitments is important because, if we're aware of them only as fully fledged intentional objects, we risk reflectively reinscribing the objectivistic distortions that Al-Saji prescribes hesitation to disrupt.

23. Note similarities with Kelly's (2004, 93) contention that, although the view-from-everywhere is unobtainable, not all situated perceptions are equal. Some, such as where the profile of a cup reveals its handle—which is intimately tied to the reason I'm ordinarily interested in cups (i.e., as drinking vessels)—are more *revealing* than others. The difference between our positions, however, again lies in the question of whether something's *most revealing* aspects might be those beyond our ordinary utilitarian purview. There is a different sort of "full bodily readiness" at work in my position, which I aim to incubate via the praxis I promote.

24. I'm not claiming that the scientists necessarily have this guidance entirely right.

CHAPTER 4: THE SPECTER OF CORRELATIONISM

1. Like Graham Harman, I understand object-oriented ontology to be a species of the genus "speculative realism." Although Tom Sparrow (2014) conscribes new materialists like Jane Bennett to the speculative realist cause, it's misleading to do so. While evidently a metaphysical realist, Bennett isn't committed to speculatively evading correlationism, nor does she necessarily concede its prevalence.

2. By adhering to an expressive theory of truth, Merleau-Pontians apparently find themselves in a worse position ("strong correlationism") than Kantians ("weak

correlationism"). Kant at least thinks that the thing-in-itself can be thought, and must exist, even if it cannot be *known* without anthropocentric filtering via the categories (Meillassoux 2008, 35).

3. Haraway seemingly shows her correlationist stripes by characterizing objectivity in terms of overlapping (human) consensus, or "webs of connections," which allow "solidarity in politics and shared conversations in epistemology" (1991, 191). Like Haraway, Husserl (1960, 106–7) contends that the world's "objectivity" may also be coconstituted by transspecies Others. Nevertheless, Meillassoux holds that, under correlationism, any nonhuman "intersubjective" contributions can be grasped only by subsuming them under the ontologically anthropocentric terms of the correlationist circle. This is because, Meillassoux claims, transcendental subjectivity is confined to a "transparent cage": the specific, incarnated, "*point of view*" in which one is "*instantiated*" and which dictates the terms of critical enquiry (2008, 7, 24–25). Although I have misgivings about Meillassoux's argument, we shall see in chapter 5 that Husserl's "transcendental conditions of experience" aren't obviously hospitable to the alterity of nonhuman body-subjects.

4. James associates these arguments for metaphysical realism with Holmes Rolston III. Rolston anticipates Meillassoux's "ancestrality" argument by referring to the "lion-objects" that must temporally precede one's identification of them (1997, 52–55). Meillassoux contends that phenomenology relies upon a world that precedes human consciousness for it to emerge. However, because of its correlationism, Meillassoux thinks, phenomenology cannot even think about the world it requires. Fully engaging with the ancestrality argument goes beyond the scope of this book. However, I soon gesture in the direction of a response via Merleau-Ponty's identification of "lateral kinship."

5. Morton claims that resisting "Western ideas of the subject as transcendence" requires reconstruing "everything as objects" and recognizing that "there is no Nature, only people, some of whom are human beings" (2008, 77; 2011, 168). Although Morton's solution to this problem differs radically from Meillassoux's, his objection is consistent with Meillassoux's lament that, through specifically *correlational* subject/object relations, "contemporary philosophers have lost the *great outdoors,* the *absolute* outside of precritical thinkers: that outside which was not relative to us, and which was given as indifferent to its own givenness to be what it is, existing in itself regardless of whether we are thinking it or not; that outside which thought could explore with the legitimate feeling of being on foreign territory—of being entirely elsewhere" (Meillassoux 2008, 7).

6. Note similarities with Merleau-Ponty's earlier discussion of the meanings available to beetles and dogs. In the human case, however, we may register a greater degree of access to the world inhabited by the body-subject in question.

7. Levinas (1993, 96–103) holds that, by understanding Others in epistemological terms, Merleau-Ponty misses the ethical moment of radical alterity. For Levinas, ethical engagement with alterity requires a "nonreciprocal relationship" (1987b, 83) between parties. However, Levinas's critique is misdirected on two counts. Levinas firstly fails to appreciate that, for Merleau-Ponty, as body-subjects who share chiasmic openings onto the world, identity and alterity become largely matters of degree rather than kind. It's straightforwardly false, therefore, to think that partial epistemic reciprocity entails, or even seeks, a relationship of objectivistic assimilation in the manner Levinas thinks. Secondly, as Jack Reynolds (2004, 315–17) argues, Levinas

also fails to appreciate that one can elicit from Merleau-Ponty's works a certain *ethical* salience to the possibility of dialogical reciprocity between parties that requires a degree of epistemic encroachment. This will prove important in what follows.

8. David Golumbia (2016, 17) emphasizes Kant's rejection of idealism in discussing the compatibility of Kantian metaphysics and realist Newtonian science, for instance. I think Harman is nonetheless more or less right that "the thing portrayed by the natural sciences is the thing made dependent on our knowledge, and not the thing in its untamed, subterranean reality" (2011b, 80). Harman and I differ, however, on whether natural science (properly situated) is idealistic.

9. There is no requirement for universal sentience, Merleau-Ponty insists, even within the context of the flesh (*VI*, 39, 136, 250).

10. Harman might also be picking up on the resonances (which Merleau-Ponty sometimes courts) with the pure plenitude of Sartrean being-in-itself. I explain why Merleau-Ponty needn't commit to a Sartrean ontology in chapter 6.

11. Merleau-Ponty warns about linguistic distortion in the terminology he employs. For Merleau-Ponty, an entity "is not first of all a meaning for the understanding, but a structure accessible to inspection by the body, and if we try to describe the real as it appears to us in perceptual experience, we find it overlaid with anthropological predicates" (*PP*, 320).

12. We should nonetheless emphasize Bennett's concerns about making such claims.

13. Shaviro blames correlationism on the bifurcation of nature into distinct objective and subjective realms (2014, 65). However, as we shall see, by collapsing subject/object distinctions *in the modality of the in-itself*, Shaviro is no less guilty of this very bifurcation.

14. Note clear parallels between the false dilemmas of empiricism/intellectualism and realism/idealism. The claim that correlationism boils down to subjective mediation of objects appears to attribute to Merleau-Ponty a misleading representationalism.

15. Merleau-Ponty makes this point repeatedly in the *Structure of Behaviour.* In its final paragraph, Merleau-Ponty argues that it is an objectivistic error to attempt to transcendentally naturalize a nonhuman Gestalt. This explains why the *Phenomenology* is so concerned with human perception—it is a phenomenological investigation carried out *within* a certain Gestalt structure—not because he abandoned nonhuman intentionality. Harman doesn't appear to be familiar with Merleau-Ponty's early work. Acceptance of nonhuman "awareness," however, might permit Merleau-Pontian ecophenomenology to pass Harman's litmus test in another way because it licenses irreducibly nonhuman expressions of the world.

16. Recall Bennett's distinction between subjectivity and quasi agency. Since Shaviro is set against "brute emergence," he needs even slime molds, thermostats, and neutrinos to possess the former. Quasi agency, however, isn't so obviously neo-Cartesian.

17. For Plumwood, master identities are also sociohistorically constructed. The atomism invoked from the master perspective, for example, is fallacious outside of the logic of dualism. This means that acts of incorporation, like Shaviro's, are shot through with misleading and unacknowledged sediment. There are more explicitly androcentric overtones in Shaviro's Baconian claim that "scientific experimentation forces [the world] to speak" (2014, 71).

18. Since object-oriented ontologists think that all objects intend or prehend one another in much the same ways, it's appropriate, if uncomfortably anthropomorphic, to refer to all objects as Others here.

19. The difference between the two positions is that Levinas never explicitly claims that we're unable to know Others because of some epistemic limitation of access. His point, rather, is that all knowing totalizes and our ethical responsibility to Others is an altogether different manner of relating. Nevertheless, to fully understand Levinas's objection to Merleau-Pontian phenomenology, he (like Harman) must think that one cannot gain even a degree of epistemic access to Others without collapsing their alterity in a colonial manner. Likewise, object-oriented ontology isn't merely concerned with epistemological or ontological matters. As Morton's work demonstrates, object-oriented ontologists ascribe an ethical salience to the attempt to address nonhuman Others on their own terms. The two positions are much closer than one might think. I explore problems with Morton's and Harman's positions further (see Forthcoming b).

20. This argument can function only at an epistemological level. The rainwater is oblivious to the vinegar's pungency, but it nevertheless (ontologically) dilutes it. The pseudo-intentional encounters object-oriented ontologists describe cannot, therefore, be "in any sense definitive" (Shaviro 2014, 105–6) of encounters more broadly.

21. Somewhat ironically, Harman (2005, 41–42) levels this criticism against Levinas.

22. If the object-oriented ontologists' claim isn't that the object stands apart, but itself somehow *motivates* asymmetrical metaphor, we're left with something—problematic atomistic ontology aside—that differs little from Merleau-Pontian "correlationism."

23. Harman even charges Merleau-Ponty with a pernicious anthropocentrism for acknowledging that one's situation makes perception *a*symmetrical (2005, 54–55). To his credit, Bogost recognizes that "anthropomorphism is unavoidable, at least for us humans," and that metaphorism is particularly guilty in this regard (2012, 64, 76). Bogost thinks, however, that anthropomorphism is a tonic for anthropocentrism. The extent to which object-oriented ontologists retain species solipsistic norms undermines Bogost's claim.

24. Vogel diagnoses a similar issue in Adorno's work, where objects retain only a "mundane" excess. This issue arises because Adorno rejects the concept of expressive truth required to take differential realities sufficiently seriously. For Adorno, expressive truth amounts to "the seemingly unbearable thesis that subjectivity presupposes facts while objectivity presupposes the subject" (Adorno 1973, 141; Vogel 1996, 73–74, 189).

25. Interestingly, Bogost (2012, 78–79) admits that second-level metaphorization takes place, at least in ethical matters, and cites Levinas as someone guilty of failing to take metaphorism seriously.

26. Perhaps the problem with advancing a blanket correlationist charge lies in assuming that "correlationists" share a common commitment to Kantian metaphysical stipulations. Although Kant disavows the experiential uncoupling of concepts and intuitions, he reifies the transcendental subject in the manner problematized by speculative realists and Merleau-Pontian ecophenomenologists alike (Kant 1996, 193–94, B75/A51).

27. To think of *whole experiences* as discursive/nondiscursive is misleading because it implies the culture/nature dualism I resist. Nevertheless, the term "nondiscursive grounds of experience" isn't entirely unproblematic. The cross-blending of human vision, for example, is fundamentally nondiscursive under these terms, but still particular to a narrow set of organisms. The world presented through binocular vision, however, remains irreducible *to* these biologically specific grounds. Thus, the

radically reflective body-subject must continue to question to what extent any particular nondiscursive grounds afford narrowly species-specific expressions. I explore this issue in chapter 5.

28. Despite often being wildly dissimilar, supernormal stimuli elicit stronger behavioral responses than the stimuli those behaviors have apparently evolved to suit. Massumi's point is that relationships of "supernormal dynamism" cannot be explained mechanistically or instinctually, since they involve a rudimentary sort of improvisation (Massumi 2015, 4).

29. In the interest of doing justice to alterity, I suggest we resist Merleau-Ponty's problematically rigid taxonomy, which is inessential to the broader claim about chiasmic kinship.

30. Toadvine (2014) uses these tools to respond to Meillassoux's ancestrality objection. If no moment of material history is entirely alien to one's perceptual milieu, then "there is no past which is absolutely past" (*SB*, 207). The "elemental" past outstrips the emergence of consciousness and any subsequent subject/object schema. The elemental past's reverberation *within* one's worldly grasp, therefore, is a condition of personal time; it doesn't *merely* occur as an object within it. Thus, one has the same ambiguous sort of continuity with the elemental past as one does with the perspectives of other animals. As is also the case with other animals, ontological continuity means that one needn't understand the distant past as a distinct object-in-itself with which one is (correlatively) hyperseparated.

31. This needn't require that each thing *has* a perspective as in Shaviro's panpsychism. Rather, where something *does*, the radically reflective body-subject may be able to acknowledge the meanings revealed in that perspective to *some* (gradated) degree. Where something *lacks* a perspective, by being attentive to shared nondiscursive grounds of experience that underpin expressions of the world, the radically reflective body-subject may again glean some degree of insight to other perspectives of that thing (as required by the view-from-everywhere).

32. Another way of reading this claim is to focus on the *possibility* of distinguishing between properties that are supposed to belong to the object rather than subjective access to that object. Read this way, speculative realism and Merleau-Pontian phenomenology *do* differ: the former may make determinate claims about which properties belong to the object-*in-itself*, the latter cannot. However, I've argued that the ability to do so either is unimportant or rests upon metaphysical stipulations that preclude appreciation of more-than-human alterity.

CHAPTER 5: ANDROCENTRISM, NONDISCURSIVE GROUNDS, AND THE HYPERDIALECTIC

1. See Jill W. Scott (1992), Luce Irigaray (1993), Judith Butler (1989), or Shannon Sullivan (1997).

2. Graham Harman (2005, 24) cites this passage in support of his cause, strengthening my contention that Harman's object-oriented ontology is insufficiently self-reflective about its (ontologically anthropocentric) sediment.

3. I hereafter refer to these interconnected group solipsisms under the banner of "androcentrism."

4. I neglect Butler's specific objection here because, like Anna P. Foultier (2013), I think Butler straightforwardly mistakes Merleau-Ponty's *descriptive* claims for hetero-*normative* ones.

5. Oksala slips between contending that there are no entirely nondiscursive *experiences* (which I accept, but which needn't entail radical antifoundationalism), to contending that there are no nondiscursive *grounds* of experience in the body schema (which I deny). There is, therefore, some ambiguity about whether she means to refer to the sort of epistemological foundationalism concerning the putative existence of a nonconceptual "given" at the level of experience (which divides Hubert Dreyfus and John McDowell), or something more fundamental about the nondiscursive limits of the body's intentional structures. Oksala is an unrepentant Butlerian. The body, for Oksala, plays more or less the same performative role as Butler's *psyche*. Oksala's ambiguity might, therefore, be explained by a mistake similar to the one identified in note 7. Nonetheless, Oksala's position *is* consistently antifoundationalist insofar as she denies foundational priority to any specific discourse, rather than discourse more broadly.

6. Alcoff's analysis should be extended to include discursive grounds which are less clearly textual (e.g., ideological power relations), but her broader point stands.

7. Like Meillassoux (2008, 27), Scott (1992, 23) is explicit that naive realism's implausibility makes knowledge claims rooted in perception irredeemably correlationist. A related error infects Butler's *metaphysical* reduction of gender to discourse on the grounds that we lack *epistemic* access to gender independent of discourse: "In the first case, Butler is arguing that gender as a concept is discursively constructed, and in the second case she is arguing that gender is discursively constructed" (Alcoff 2006, 167). Butler may be right that gender isn't a natural kind. However, Butler fails to take seriously the material features of bodies which really may "cluster," as Alison Stone (2008, 155) puts it.

8. Minimally: the possession of pain receptors and certain neurological capacities. I'm wary, however, of the scientistic reductionism involved in neglecting to situate biological features within the context of embodied intentionality. The following may also be relevant: the self-givenness of (relative) bodily integrity, the experiential primacy of phenomenal ("personal") space over "objective" space, and the operative *capacity* to experience the meaning of danger *as* danger. While these intentional norms might seem unduly sophisticated for an unremarkable juvenile body-subject, we should be reminded that they seem to be enjoyed by even plant life, to some degree.

9. "4E" stands for "embodied," "embedded," "extended," and "enactive" approaches (see, e.g., Gibson 1979; Gallagher 2005; or Clark 2008). I touch upon one such dialectic in this chapter (5.4.1.n17), and one in the next (6.1.1).

10. Kruks would think that Fanon misunderstands his participation in the master/slave dialectic. As embodied, the master *must* recognize the slave to some degree as an alter ego. Nevertheless, in cases like Fanon's, the master's pathological response to this inassimilable event is to objectify the slave.

11. I'm grateful to an anonymous reviewer of an earlier draft who, in evidence of this point, suggested that the violent language I had sometimes (unwittingly) used to characterize my relationships with some other thinkers may be unproductive to the book's collaborative aspirations.

12. As Fischer notes, even Irigaray requires a common "mucosity" to sustain intersubjective relations between sexuate beings and, like Merleau-Ponty, focuses upon the role of dialectical denial (what Irigaray calls the "negative") in facilitating openness to alterity (Fischer 2007, 208–9; Irigaray 1996).

13. They are both "middle-class" and human, for instance.

14. I focus here on alternative phenomenologies and the possibility of one's openness to them. However, as Jack Reynolds (2016) argues, we also find "fecund contradictions" in interdisciplinary dialectics. One fruitful dialectic emerges between Merleau-Ponty and James J. Gibson's (1966; 1979) work in cognitive science. We've seen that Merleau-Ponty used Gestalt psychology to challenge the phenomenological givenness of Husserl's transcendental essences. Gibson informed the phenomenological debate by exposing dated assumptions about the reductive presumptions of all alternative sensory psychologies. Although Gibson sees his work as a continuation of Merleau-Ponty's, Gibson's environmental approach might nevertheless concede too much to the objectivistic remit of classical cognitive science through its homogenizing appeal to affordances. To explain: by "affordances" Gibson means the possibilities that an environment provides for an organism relative to its functional orientation, intentions, and so on. Gibson's affordances are inherently relational and, therefore, cannot be described in atomistic terms without distortion. Nonetheless, Gibson takes affordances to be independent of any given animal's ability to encounter them as such. As cognitive scientists, Gibson appears to think, *we* may identify *the* affordances, even if the earthworm misses them. Thus, not unlike John Odling-Smee, Gibson ends up naturalizing the phenomena somewhat specific to a certain type of Gestalt (i.e., the affordances "objectively" available for detection) from an apparently transcendent standpoint. Gibson thereby demonstrates a problematic hostility to nonhuman alterity. Meacham notes that a similarly problematic ontological anthropocentrism pervades the bioenactivist literature that Gibson's research spawned, since "many of its operative concepts seem to be drawn, at least initially, from philosophy of mind and human phenomenology and then scaled down to less complex systems" (Meacham 2016, 75). Merleau-Pontians, however, are better disposed to attend to this issue because "Merleau-Ponty, cognizant of his own Umwelt-situatedness, maintains that any access to this virtuality, what he calls brute or wild being, must be through perception" (Meacham 2016, 92).

Reynolds takes dialectics like the above to show Merleau-Ponty to be a weak methodological naturalist (i.e., to hold "that philosophical results ought to be broadly continuous with those of the sciences" [Reynolds 2016, 83]). I deny this claim for two reasons. First, because Merleau-Ponty (rightly) affords a certain primacy to phenomenology over derivative scientific revelations. Weak methodological naturalists more or less affirm the inverse. Second, because, as Reynolds (2016, 102) himself notes, weak methodological naturalists attempt to *eliminate* interdisciplinary contradictions between findings. I've argued, however, that sustaining the contradictions viewed with suspicion by naturalists (e.g., via a view-from-everywhere) may permit greater openness to more-than-human alterity. My interpretation coheres better with Heinämaa's than Reynolds's.

CHAPTER 6: RADICAL REFLECTION, REVERSIBILITY, AND THE FLESH

1. This is a tricky task since *The Visible and the Invisible* remained unfinished in Merleau-Ponty's lifetime. I explore what I take to be the most compelling interpretation.

2. Merleau-Ponty himself thought a Sartrean ontology followed from our freedom to negate our "vital situation" (*PP*, 400) via reflective consciousness. However,

Toadvine (2009, 73–75) notes, Merleau-Ponty was merely acting under the influence of Max Scheler, whose philosophy Merleau-Ponty *rejects* because of its sharp nature/consciousness binary.

3. Some ecophenomenologists (e.g., Jensen 2007) use arguments from this text to justify a deep ecological ethic that (mistakenly) draws warrant from the undifferentiated fleshy kinship one apparently shares with all bodies.

4. I leave aside whether Being and flesh are synonymous.

5. This interpretation chimes with Robert Vallier's (2009).

6. Plumwood (2001, 12–16) contends that deep ecology is hostile to alterity because it invokes one or more of the following views of selfhood. The "indistinguishable self" viewpoint strictly *identifies* "self" with "nature" and, therefore, self-interest with the interests of the whole biotic web or cosmos. Intimately related to the homogenizing eco-holistic schemas we criticized in John Baird Callicott, Warwick Fox, and Paul Crutzen, it is incompatible with appreciating something's differential needs or wellbeing where these cannot be equated with one's own. The "expanded self" viewpoint (which Plumwood associates with Arne Naess) demands *empathetic* identification with the totality of worldly entities. It is problematic insofar as it cannot critique the logic of egoism, but rather expands it without reserve. The "transpersonal self" viewpoint requires an *impartial* identification of "self"—reflectively purified of interests and commitments—with all other aspects of the "natural" world. Hostile to the *particularity* of individuals and their local responsibilities and needs, it is nonaccidentally aligned with tendencies toward homogenization of the ilk we criticized in Peter Singer and Callicott.

7. Remember that the *Phenomenology*'s descriptions of experience haven't been dismissed in Merleau-Ponty's later work; he's trying to better explain them.

8. Note that this objection would only be fatal for the position advanced in the *Phenomenology* if we mistook Merleau-Ponty's approach to be necessarily foundationalist.

9. We shouldn't, however, neglect Irigaray's objection to the point Bannon raises about the structural homogeneity of perception apparently experienced by all sentient creatures.

10. "Pure reversibility" refers to the claim that one can more or less transparently take up the other party's position. While the touched/touching experiences of each hand have minor qualitative differences, there remains an *imminent* pure reversibility since one can transparently imagine what the other side would be like. It's misleading to use the touching of one's own hands as an *exemplar sensible,* however, since, being part of the same body, one's hands *can* oscillate between touching and touched in a manner untrue of intersubjective or interobjective experiences. Levinas (1993, 96–103) makes a similar point.

11. Merleau-Ponty's conclusion that one's *experience* "does not notably differ from that of the others" now seems overdrawn. However, this claim *might* have been tempered in the final editing process.

12. See Brook (2005, 361).

13. So long as these neo-Cartesians don't maintain an *exclusive* and *exhaustive* claim to truth (which they ordinarily do).

14. The flesh doesn't *temporally* predate subject and object. A "psychoanalysis of Nature" reveals the flesh's prethematic unity, in a state of noncoincidence, through the particular subject and object poles it envelops. "A philosophy of flesh," Merleau-Ponty claims, is therefore "the condition without which psychoanalysis remains

anthropology" (*VI*, 267). The yardstick problem, however, may undermine this promise.

15. Barbaras's (2004, 169, 202–3) suggestion that the flesh extends outward only from the human body must be incorrect. It would threaten the connaturality of entities by allowing things to *become* flesh only through human perception (Bannon 2011, 341). Merleau-Ponty stresses that the flesh cannot be limited to its human manifestations (*N*, 216–17; *VI*, 172). By allowing nonhuman entities primary participation within it, Toadvine's interpretation becomes more plausible than Barbaras's. A related problem with Toadvine's interpretation arises, however, *if* human language is the *only* means by which to express "what the world means to say."

16. For Heidegger (1969, 42–74), "ontotheology" is the emblematic error of "Western" metaphysics. It effectively licenses a project of anthropocentric mastery by running together ontological and theological discourses, subsuming their respective mysteries and excesses into wholly intelligible metaphysical systems and objects to which humans have privileged access.

17. Although, if all is flesh, such expressions possess a curious ontological status.

18. If these schemas rely unambiguously on (e.g., totalizing substance dualist) ontologies straightforwardly inconsistent with the flesh, then they would presumably be ruled out by someone who adheres to it. Nonetheless, I'm not sure that many contemporary proponents of the problematic purviews explored in this book assert ontological claims which are *straightforwardly* inconsistent with the flesh.

19. Given their more nebulous or intangible nature, our issue may grow in seriousness where axiological meanings are concerned.

20. Barbaras's contention that ontology's absolute anthropomorphism is "simply a truth with which we must reckon" won't much benefit the ecophenomenological projects in which the flesh is usually employed (Barbaras 2002, 22–25; Bannon 2011, 335).

21. E.g., Toadvine (2009, 133); Brook (2005, 361); Fischer (2007).

22. Indeed, insofar as these critical tools rely upon Bannon's specific (i.e., radically ecological) interpretation of the flesh, they may also clash with some specifically phenomenological insights previously discussed (e.g., Gallagher and Meltzoff's contentions about the primordial self).

23. Vogel thinks it is *impossible* to raise the question of accuracy where nonhuman meanings or contributions are concerned. I've argued that Vogel's claim is too strong. He nevertheless foregrounds an inherent asymmetry in dialectical relationships which appeals to the flesh, partly because claims about what the world "wants" to say are fundamentally unanalyzable, are apt to mask.

24. Bannon and Toadvine, for instance, are led by Merleau-Ponty's (somewhat unconvincing) arguments that the flesh alone prevents phenomenology from becoming a philosophy of consciousness. Toadvine (2009, 133) then takes the motor of *all* expressive meanings to be "nature" itself, which explains why he doesn't think the flesh straightforwardly licenses any specific maxim. Bannon (2011, 333–35, 352–55) thinks that Toadvine's radicalization of (human) phenomenology anthropomorphizes the flesh, arguing that nonanthropomorphism requires we understand it as a broader scheme of generalized affective relations.

25. It may be true that any fixed ontology limits its terms of critical investigation. So, perhaps the point (at least for people like Toadvine) is this: *if* one wishes to advance a fundamental ontology, then the flesh seems to be more congenial to doing critical environmental philosophy than most. However, I've argued that one may do

critical environmental philosophy just as effectively, and probably more so, without invoking a fundamental ontology which carries the extra baggage the flesh brings.

"CONCLUSIONS"; OR, BECOMING A PLACE OF UNREST

1. Note my emphasis on the "in-themselves" clause; I don't problematize the reality of meanings, nor various normative grounds thereof, as such.

2. While the first may be a genuine paradox, attention to different senses of "anthropocentrism" suggests the second to be a merely *apparent* paradox.

REFERENCES

Abram, David. 1996. *The Spell of the Sensuous: Perception and Language in a More-Than-Human World*. New York: Vintage Books.

Adorno, Theodor W. 1973. *Negative Dialectics*. Translated by E. B. Ashton. New York: Continuum.

Alaimo, Stacy. 2007. "Trans-corporeal Feminisms and the Ethical Space of Nature." In *Material Feminisms*, edited by Stacy Alaimo and Susan Hekman, 237–64. Bloomington: Indiana University Press.

———. 2014. "Thinking as the Stuff of the World." *O-Zone: A Journal of Object-Oriented Studies* 1 (1): 13–21.

———. 2017. "When the Newt Shut Off the Lights: Scale, Practice, Politics." In *Teaching Climate Change in the Humanities*, edited by Stephen Siperstein, Shane Hall, and Stephanie LeMenager, 31–36. London: Routledge.

Alcoff, Linda Martín. 2000. "Merleau-Ponty and Feminist Theory on Experience." In *Chiasms: Merleau-Ponty's Notion of Flesh*, edited by Fred Evans and Leonard Lawlor, 251–72. Albany, NY: SUNY Press.

———. 2006. *Visible Identities: Race, Gender and the Self*. Oxford: Oxford University Press.

———. 2010. "Epistemic Identities." *Episteme* 7 (2): 128–37.

Alpi, Amedeo, Nikolaus Amrhein, Adam Bertl, Michael R. Blatt, Eduardo Blumwald, Felice Cervone, Jack Dainty, Maria Ida De Michelis, Emanuel Epstein, Arthur W. Galston, Mary Helen M. Goldsmith, Chris Hawes, Rüdiger Hell, Alistair Hetherington, Herman Hofte, Gerd Juergens, Chris J. Leaver, Anna Moroni, Angus Murphy, Karl Oparka, Pierdomenico Perata, Hartmut Quader, Thomas Rausch, Christophe Ritzenthaler, Alberto Rivetta, David G. Robinson, Dale Sanders, Ben Scheres, Karin Schumacher, Hervé Sentenac, Clifford L. Slayman, Carlo Soave, Chris Somerville, Lincoln Taiz, Gerhard Thiel, and Richard Wagner. 2007. "Plant Neurobiology: No Brain, No Gain?" *Trends in Plant Science* 12 (4): 135–36.

Al-Saji, Alia. 2009. "A Phenomenology of Critical-Ethical Vision: Merleau-Ponty, Bergson, and the Question of Seeing Differently.' *Chiasmi International* 11:375–98.

Angelin, Sallie. 2015. "Generative Motion: Queer Ecology and Avatar." *Journal of Popular Culture* 48 (2): 341–54.

Arias-Maldonado, Manuel. 2016. *Environment and Society: Socionatural Relations in the Anthropocene*. Dordrecht: Springer.

Arthur, Charles. 1999. "Scientists Discover Why Fungi Have 36,000 Sexes." *Independent*, September 14, 1999. http://www.independent.co.uk/news/scientists-discover-why-fungi-have-36000-sexes-1119181.html.

Awuah-Nyameky, Samuel. 2014. *Managing the Environmental Crisis in Ghana: The Role of African Traditional Religion and Culture with Special Reference to the Berekum Traditional Area*. Newcastle upon Tyne: Cambridge Scholars Publishing.

Bannon, Bryan E. 2009. "Developing Val Plumwood's Dialogical Ethical Ontology and Its Consequences for a Place-Based Ethic." *Ethics and the Environment* 14 (2): 39–55.

———. 2011. "Flesh and Nature: Understanding Merleau-Ponty's Relational Ontology." *Research in Phenomenology* 41 (3): 327–57.

———. 2014. *From Mastery to Mystery: A Phenomenological Foundation for an Environmental Ethic*. Athens: Ohio University Press.

Barad, Karen. 2006. *Meeting the Universe Halfway: Quantum Physics and the Entanglement of Matter and Meaning*. Durham, NC: Duke University Press.

———. 2008. "Posthumanist Performativity: Toward an Understanding of How Matter Comes to Matter." In *Material Feminisms*, edited by Stacy Alaimo and Susan Hekman, 120–54. Bloomington: Indiana University Press.

Barbaras, Renaud. 2002. "The Ambiguity of the Flesh." *Chiasmi International* 4:13–26.

———. 2004. *The Being of the Phenomenon: Merleau-Ponty's Ontology*. Translated by Ted Toadvine and Leonard Lawlor. Bloomington: Indiana University Press.

Bartky, Sandra. 1990. *Femininity and Domination: Studies in the Phenomenology of Oppression*. London: Routledge.

BBC. 2017. "Paris Climate Deal: Trump Pulls US out of 2015 Accord." *BBC*, June 1, 2017. https://www.bbc.co.uk/news/world-us-canada-40127326.

Beauvoir, Simone de. 1952. *The Second Sex*. Translated and edited by H. M. Parshley. New York: Vintage Books.

Bennett, Jane. 2010. *Vibrant Matter: A Political Ecology of Things*. Durham, NC: Duke University Press.

———. 2011. "Response to Thomas Princen's Review of *Vibrant Matter: A Political Ecology of Things*." *Perspectives on Politics* 9 (1): 120–21.

Bernet, Rudolf. 1993. "The Subject in Nature: Reflections on Merleau-Ponty's Philosophy of Nature." In *Merleau-Ponty in Contemporary Perspectives*, edited by Philip and Jan van der Veken, 53–68. Dordrecht: Kluwer.

Blackless, Melanie, Anthony Charuvastra, Amanda Derryck Castel, Anne Fausto-Sterling, Karl Lauzanne, and Ellen Lee. 2000. "How Sexually Dimorphic Are We? Review and Synthesis." *American Journal of Human Biology* 12 (2): 151–66.

Bleischwitz, Raimund, Monika Dittrich, and Chiara Pierdicca. 2015. "Illicit Trade with Coltan and Implications for Certification." In *Competition and Conflicts on Resource Use*, edited by Susanne Hartard and Wolfgang Liebert, 155–75. Dordrecht: Springer.

Bogost, Ian. 2012. *Alien Phenomenology, or What It's Like to Be a Thing*. Minneapolis: University of Minnesota Press.

Booth, Robert. Forthcoming a. "Process Metaphysics and the Shadow of Kant." In *ABC+ Process Philosophy*, edited by Robert Booth and Oliver George Downing. Berlin: De Gruyter.

———. Forthcoming b. "Abject Withdrawal? On the Prospect of a Nonanthropocentric Object-Oriented Ontology." *Angelaki* 26 (5).

Bordo, Susan. 1987. *The Flight to Objectivity: Essays on Cartesianism and Culture*. Albany, NY: SUNY Press.

Boyle, Louise. 2020. "Himalayas Seen for First Time in Decades from 125 Miles Away after Pollution Drop." *The Independent*, April 8, 2020. https://www.independent.co.uk/environment/himalayas-mountain-range-asia-india-pollution-coronavirus-lockdown-a9456446.html.

Brassier, Ray. 2007. *Nihil Unbound: Enlightenment and Extinction*. Basingstoke, UK: Palgrave Macmillan.

Brook, Isis. 2005. "Can Merleau-Ponty's Notion of 'Flesh' Inform or Even Transform Environmental Thinking?" *Environmental Values* 14 (3): 353–62.

Brown, Charles S. 2003. "The Real and the Good: Phenomenology and the Possibility of an Axiological Rationality." In *Eco-phenomenology: Back to the Earth Itself*, edited by Charles S. Brown and Ted Toadvine, 3–18. Albany, NY: SUNY Press.

Brown, Lawrence. 2015. "The Ghost of King Leopold II Still Haunts Us: Belgium Colonization and the Ignition of the HIV Global Pandemic." *Media Diversified*, April 20, 2015. https://mediadiversified.org/2015/04/20/the-ghost-of-king-leopold-ii-still-haunts-us-belgium-colonization-the-ignition-of-the-hiv-global-pandemic/.

Burns, Kelly. 2008. "Warren's Ecofeminist Ethics and Merleau-Ponty's Body Subject." *Ethics and the Environment* 13 (2): 101–18.

Butler, Judith. 1989. "Sexual Ideology and Phenomenological Description." In *The Thinking Muse: Feminism and Modern French Philosophy*, edited by Jeffner Allen and Iris Marion Young, 85–100. Bloomington: Indiana University Press.

Callicott, John Baird. 1989. *In Defense of the Land Ethic: Essays in Environmental Philosophy*. Albany, NY: SUNY Press.

———. 1999. *Beyond the Land Ethic: More Essays in Environmental Philosophy*. Albany, NY: SUNY Press.

Calvert-Minor, Chris. 2013. "Epistemological Misgivings of Karen Barad's Posthumanism." *Human Studies* 37 (1): 123–37.

Carman, Taylor. 2005. "Sensation, Judgment, and the Phenomenal Field." In *The Cambridge Companion to Merleau-Ponty*, edited by Taylor Carman and Mark B. N. Hansen, 50–73. Cambridge: Cambridge University Press.

Carrington, David. 2016. "The Anthropocene Epoch: Scientists Declare Dawn of Human-Influenced Age." *The Guardian*, August 29, 2016. https://www.theguardian.com/environment/2016/aug/29/declare-anthropocene-epoch-experts-urge-geological-congress-human-impact-earth.

Casey, Edward S. 2001. "Taking a Glance at the Environment: Prolegomena to an Ethics of the Environment." *Research in Phenomenology* 31 (1): 1–21.

Cerbone, David R. 2008. "Perception." In *Merleau-Ponty: Key Concepts*, edited by Rosalyn Diprose and Jack Reynolds, 121–31. Stocksfield, UK: Acumen.

Clark, Andy. 2008. *Supersizing the Mind: Embodiment, Action and Cognitive Extension*. Oxford: Oxford University Press.

Clarke, Melissa. 2002. "Ontology, Ethics, and *Sentir*: Properly Situating Merleau-Ponty." *Environmental Values* 11 (2): 211–25.

Coleman, Sam. 2006. "Being Realistic—Why Physicalism May Entail Panexperientialism." *Journal of Consciousness Studies* 13 (10–11): 40–52.

Cook, Deborah. 2011. *Adorno on Nature*. Stocksfield, UK: Acumen.
Cronon, William. 1995. "The Trouble with Wilderness; or, Getting Back to the Wrong Nature." In *Uncommon Ground: Rethinking the Human Place in Nature*, edited by William Cronon, 69–90. New York: W. W. Norton.
Crutzen, Paul. 2002. "Geology of Mankind." *Nature* 415: 23.
Crutzen, Paul, and Christian Schwägerl. 2011. "Living in the Anthropocene: Toward a New Global Ethos." *Yale Environment 360*, April 7, 2017. http://e360.yale.edu/features/living_in_the_anthropocene_toward_a_new_global_ethos.
Cuomo, Chris J. 2017. "Against the Idea of an Anthropocene Epoch: Ethical, Political and Scientific Concerns." *Biogeosystem Technique* 4 (1): 4–8.
Dainton, Barry. 2008. *The Phenomenal Self*. Oxford: Oxford University Press.
DeLanda, Manuel. 1995. "Uniformity and Variability: An Essay in the Philosophy of Matter." http://www.to.or.at/delanda/matterdl.htm.
———. 2000. *A Thousand Years of Nonlinear History*. New York: Zone Books.
Deleuze, Gilles, and Félix Guattari. 2004. *A Thousand Plateaus: Capitalism and Schizophrenia*. Translated by Brian Massumi. London: Continuum.
Descartes, René. 1996. *Meditations on First Philosophy*. Translated by John Cottingham. Cambridge: Cambridge University Press.
Dillon, M. C. 1988. *Merleau-Ponty's Ontology*. Evanston, IL: Northwestern University Press.
Douglas, Heather. 2016. "Values in Science." In *The Oxford Handbook of Philosophy of Science*, edited by Paul Humphreys, 609–32. Oxford: Oxford University Press.
Duhem, Pierre. 1982. *The Aim and Structure of Physical Theory*. Princeton, NJ: Princeton University Press.
Ellis, Erle. 2011. "Anthropogenic Transformation of the Terrestrial Biosphere." *Philosophical Transactions of the Royal Society A* 369 (1968): 1010–35.
Evernden, Neil. 1993. *The Natural Alien*. 2nd ed. Toronto: University of Toronto Press.
Fanon, Frantz. 2008. *Black Skin, White Masks*. New ed. Translated by Charles Lam Markmann. London: Pluto Press.
Fischer, Sally. 2007. "Social Ecology and the Flesh: Merleau-Ponty, Irigaray, and Eco-communitarian Politics." In *Merleau-Ponty and Environmental Philosophy: Dwelling on the Landscapes of Thought*, edited by Suzanne L. Cataldi and William S. Hamrick, 206–16. Albany, NY: SUNY Press.
Foltz, Bruce, and Robert Frodeman. 2004. "Introduction: The Nature of Environmental Philosophy." In *Rethinking Nature: Essays in Environmental Philosophy*, edited by Bruce V. Foltz and Robert Frodeman, 1–12. Bloomington: Indiana University Press.
Foucault, Michel. 1977. "Theatrum Philosophicum." In *Language, Counter-memory, Practice*, translated and edited by Donald F. Bouchard, 165–98. Ithaca, NY: Cornell University Press.
———. 1991. *The History of Sexuality*. Vol. 1, *An Introduction*. Translated by Robert Hurley. London: Penguin.
Foultier, Anna P. 2013. "Language and the Gendered Body: Butler's Early Reading of Merleau-Ponty." *Hypatia* 28 (4): 767–83.
Fox, Warwick. 1990. *Toward a Transpersonal Ecology: Developing New Foundations for Environmentalism*. Boston: Shambhala.

———. 1995. "The Deep Ecology-Ecofeminism Debate and Its Parallels." In *Deep Ecology for the Twenty-First Century*, edited by George Sessions, 5–25. Boston: Shambhala.

Frye, Marilyn. 1983. *The Politics of Reality: Essays in Feminist Theory*. Trumansburg, NY: Crossing Press.

Gaard, Greta. 1997. "Toward a Queer Ecofeminism." *Hypatia* 12 (1): 114–37.

———. 2015. "Ecofeminism and Climate Change." *Women's Studies International Forum* 49 (March–April): 20–33.

Gallagher, Shaun. 2005. *How the Body Shapes the Mind*. Oxford: Oxford University Press.

Gallagher, Shaun, and Andrew N. Meltzoff. 1996. "The Earliest Sense of Self and Others: Merleau-Ponty and Recent Developmental Studies." *Philosophical Psychology* 9 (2): 211–33.

Gibson, James J. 1966. *The Senses Considered as Perceptual Systems*. Boston: Houghton Mifflin.

———. 1979. *The Ecological Approach to Visual Perception*. Boston: Houghton Mifflin.

Glazebrook, Trish. 2002. "Karen Warren's Ecofeminism." *Ethics and the Environment* 7 (2): 12–26.

———. 2005. "Gynocentric Eco-Logics." *Ethics and the Environment* 10 (2): 78–99.

———. 2011. "Women and Climate Change: A Case-Study from Northeast Ghana." *Hypatia* 26 (4): 762–82.

Golumbia, David. 2016. "Correlationism: The Dogma That Never Was." *Boundary 2* 43 (2): 1–25.

Gramsci, Antonio. 1971. *Selections from the Prison Notebooks of Antonio Gramsci*. Edited and translated by Quintin Hoare and Geoffrey Nowell Smith. New York: International Publishers.

Grimshaw, Jean. 1999. "Working Out with Merleau-Ponty." In *Women's Bodies: Discipline and Transgression*, edited by Jane Arthurs and Jean Grimshaw, 91–116. London: Cassell.

Grosz, Elizabeth. 2005. *Time Travels: Feminism, Nature, Power*. Crow's Nest, Australia: Allen and Unwin.

Hailwood, Simon. 2016. "Anthropocene: Delusion, Celebration and Concern." In *Environmental Politics and Governance in the Anthropocene: Institutions and Legitimacy in a Complex World*, edited by Philipp Pattberg and Fariborz Zelli, 47–61. London: Routledge.

Hamwey, Robert. 2020. "Environmental Impacts of Coronavirus Crisis, Challenges Ahead." *United Nations Conference on Trade and Development*, April 20, 2020. https://unctad.org/en/pages/newsdetails.aspx?OriginalVersionID=2333.

Haraway, Donna J. 1989. *Primate Visions: Gender, Race, and Nature in the World of Modern Science*. London: Routledge.

———. 1991. *Simians, Cyborgs, and Women: The Reinvention of Nature*. London: Routledge.

Harding, Sandra. 1998. *Is Science Multicultural? Postcolonialisms, Feminisms, and Epistemologies*. Bloomington: Indiana University Press.

Harman, Graham. 2005. *Guerrilla Metaphysics: Phenomenology and the Carpentry of Things*. Chicago: Open Court Press.

———. 2009. *Prince of Networks: Bruno Latour and Metaphysics*. Melbourne: Re.Press.

———. 2011a. *Quentin Meillassoux: Philosophy in the Making*. Edinburgh: Edinburgh University Press.

———. 2011b. *The Quadruple Object*. Alresford, UK: Zero Books.

Hawking, Stephen, and Leonard Mlodinow. 2010. *The Grand Design*. London: Bantam.

Hayward, Tim. 1997. "Anthropocentrism: A Misunderstood Problem." *Environmental Values* 6 (1): 49–63.

Hegel, Georg W. F. 1977. *Phenomenology of Spirit*. Translated by A. V. Miller. Oxford: Oxford University Press.

Heidegger, Martin. 1962. *Being and Time*. Translated by John Macquarrie and Edward Robinson. Oxford: Blackwell.

———. 1977. *The Question Concerning Technology and Other Essays*. Translated by William Lovitt. New York: Harper and Row.

———. 1996. *Basic Writings*. Edited by David F. Krell. Oxford: Blackwell.

———. 1999. *Contributions to Philosophy (From Enowning)*. Translated by Parvis Emad and Kenneth Maly. Bloomington: Indiana University Press.

Heinämaa, Sara. 1999. "Simone de Beauvoir's Phenomenology of Sexual Difference." *Hypatia* 14 (4): 114–32.

———. 2009. "Phenomenological Responses to Gestalt Psychology." In *Psychology and Philosophy: Inquiries into the Soul from Late Scholasticism to Contemporary Thought*, edited by Sara Heinämaa and Martina Reuter, 263–84. Dordrecht: Springer.

———. 2018. "Embodiment and Bodily Becoming." In *The Oxford Handbook of the History of Phenomenology*, edited by Dan Zahavi, 533–58. Oxford: Oxford University Press.

Hintikka, Jakko, and Merrill B. Hintikka. 1983. "How Can Language Be Sexist?" In *Discovering Reality: Feminist Perspectives on Epistemology, Metaphysics, Methodology, and Philosophy of Science*, edited by Sandra Harding and Merrill B. Hintikka, 139–48. Dordrecht: Kluwer.

Husserl, Edmund. 1960. *Cartesian Meditations: An Introduction to Phenomenology*. Translated by Dorion Cairns. Dordrecht: Springer.

———. 1970. *The Crisis of European Sciences and Transcendental Phenomenology: An Introduction to Phenomenological Philosophy*. Translated by David Carr. Evanston, IL: Northwestern University Press.

———. 2001. *Logical Investigations*. Vol. 2. Translated by J. N. Findlay. Edited by Dermot Moran. London: Routledge.

Ihde, Don. 1993. *Philosophy of Technology: An Introduction*. New York: Paragon House.

Institute for Public Policy Research. 2020. "Faster, Further, Fairer: Putting People at the Heart of Tackling the Climate and Nature Emergency." May 2020. https://www.ippr.org/files/2020-07/faster-further-fairer-ejc-interim-may20.pdf.

Intergovernmental Panel on Climate Change. 2018. "Global Warming of 1.5°C." October 8, 2018. https://www.ipcc.ch/sr15/.

Irigaray, Luce. 1993. *An Ethics of Sexual Difference*. Translated by Carolyn Burke and Gillian C. Gill. Ithaca, NY: Cornell University Press.

———. 1996. *I Love to You: Sketch of a Possible Felicity in History*. Translated by Alison Martin. London: Routledge.

Jäkel, Frank, Manish Singh, Felix A. Wichmann, and Michael H. Herzog. 2016. "An Overview of Quantitative Approaches in Gestalt Perception." *Vision Research* 126 (September): 3–8.

James, Simon P. 2007. "Merleau-Ponty, Metaphysical Realism and the Natural World." *International Journal of Philosophical Studies* 15 (4): 501–19.

———. 2009. *The Presence of Nature: A Study in Phenomenology and Environmental Philosophy*. Basingstoke, UK: Palgrave Macmillan.

Jax, Kurt, David N. Barton, Kai M.A. Chan, Rudolf de Groot, Ulrike Doyle, Uta Eser, Christoph Görg, Erik Gómez-Baggethun, Yuliana Griewald, Wolfgang Haber, Roy Haines-Young, Ulrich Heink, Thomas Jahn, Hans Joosten, Lilin Kerschbaumer, Horst Korn, Gary W. Luck, Bettina Matzdorf, Barbara Muraca, Carsten Neßhöver, Brian Norton, Konrad Ott, Marion Potschin, Felix Rauschmeyer, Christina V. Haaren, and Sabine Wichmann. 2013. "Ecosystem Services and Ethics." *Ecological Economics* 93 (C): 260–68.

Jensen, Molly Hadley. 2002. "Fleshing Out a Relational Ethics: Maurice Merleau-Ponty's Contributions to Ecological Feminism." PhD diss., Vanderbilt University.

———. 2007. "Fleshing Out an Ethic of Diversity." In *Merleau-Ponty and Environmental Philosophy: Dwelling on the Landscapes of Thought,* edited by Suzanne L. Cataldi and William S. Hamrick, 191–202. Albany, NY: SUNY Press.

Jensen, Rasmus Thybo. 2013. "Merleau-Ponty and the Transcendental Problem of Bodily Agency." In *The Phenomenology of Embodied Subjectivity*, edited by Rasmus Thybo Jensen and Dermot Moran, 43–62. Dordrecht: Springer.

Johnson, Greg. 2008. "Merleau-Ponty, Reciprocity and the Reversibility of Perspectives." In *Intertwinings: Interdisciplinary Encounters with Merleau-Ponty*, edited by Gail Weiss, 169–87. Albany, NY: SUNY Press.

Kahan, Dan. 2016. "The Politically Motivated Reasoning Paradigm, Part 1: What Politically Motivated Reasoning Is and How to Measure It." In *Emerging Trends in the Social and Behavioral Sciences*, edited by Robert A. Scott and Stephen Michael Kosslyn, 1–16. Hoboken, NJ: Wiley.

Kant, Immanuel. 1996. *Critique of Pure Reason* (unified edition). Translated by W. Pluhar. Indianapolis: Hackett.

Kelly, Sean. 2004. "Seeing Things in Merleau-Ponty." In *The Cambridge Companion to Merleau-Ponty*, edited by Taylor Carman and Mark B. N. Hansen, 74–111. Cambridge: Cambridge University Press.

Koffka, Kurt. 1925. *The Growth of the Mind: An Introduction to Child Psychology.* Translated by Robert M. Ogden. New York: Harcourt Brace.

Kohák, Erazim. 1984. *The Embers and the Stars: A Philosophical Inquiry into the Moral Sense of Nature*. London: University of Chicago Press.

Kojève, Alexandre. 1980. *Introduction to the Reading of Hegel: Lectures on the Phenomenology of Spirit*. Translated by James H. Nichols, Jr. Edited by Allan Bloom. Ithaca, NY: Cornell University Press.

Korsgaard, Christine M. 2005. "Fellow Creatures: Kantian Ethics and Our Duties to Animals." In *The Tanner Lectures on Human Values,* vol. 25, edited by G. Peterson, 79–110. Salt Lake City: University of Utah Press.

Kruks, Sonia. 2006. "Merleau-Ponty and the Problem of Difference in Feminism." In *Feminist Interpretations of Maurice Merleau-Ponty*, edited by Dorothea Olkowski and Gail Weiss, 25–48. University Park: Pennsylvania State University Press.

Kuhn, Thomas S. 1970. *The Structure of Scientific Revolutions.* 2nd ed. Chicago: University of Chicago Press.

Lacey, Hugh. 2005. *Values and Objectivity in Science: The Current Controversy about Transgenic Crops.* Lanham, MD: Lexington Books.

Langer, Monika. 1990. "Merleau-Ponty and Deep Ecology." In *Ontology and Alterity in Merleau-Ponty*, edited by Galen A. Johnson and Michael B. Smith, 115–29. Evanston, IL: Northwestern University Press.

Latour, Bruno. 1993. *We Have Never Been Modern.* Translated by Catherine Porter. Cambridge, MA: Harvard University Press.

———. 1999. *Pandora's Hope: Essays on the Reality of Science Studies.* Cambridge, MA: Harvard University Press.

Levi, Isaac. 1960. "Must the Scientist Make Value Judgements?" *Journal of Philosophy* 57 (May): 345–57.

Levinas, Emmanuel. 1987a. "Infinity and the Idea of Philosophy." In *Collected Philosophical Papers*, translated by Alphonso Lingis, 47–60. Dordrecht: Martinus Nijhoff.

———. 1987b. *Time and the Other.* Translated by Richard Cohen. Pittsburgh: Duquesne University Press.

———. 1993. *Outside the Subject.* Translated by Michael B. Smith. Stanford, CA: Stanford University Press.

Lewis, David. 1966. "An Argument for the Identity Theory." *Journal of Philosophy* 63 (1): 17–25.

Lewontin, Richard. 2000. *The Triple Helix: Gene, Organism and Environment.* London: Harvard University Press.

Li, Huey-li. 1993. "A Cross-cultural Critique of Ecofeminism." In *Ecofeminism: Women, Animals, Nature*, edited by Greta Gaard, 272–94. Philadelphia: Temple University Press.

Longino, Helen. 1996. "Cognitive and Non-cognitive Values in Science: Rethinking the Dichotomy." In *Feminism, Science, and the Philosophy of Science*, edited by Lynn Hankinson Nelson and Jack Nelson, 39–58. Dordrecht: Kluwer.

Lorde, Audre. 1983. "The Master's Tools Will Never Dismantle the Master's House." In *This Bridge Called My Back: Writings by Radical Women of Color*, edited by Cherríe Moraga and Gloria Anzaldúa, 94–101. New York: Kitchen Table Press.

Madison, Gary. 2001. "The Ethics and Politics of the Flesh." In *Merleau-Ponty's Later Works and Their Practical Implications*, edited by Duane H. Davis, 161–85. New York: Humanities Books.

Mahendran, Dilan. 2007. "The Facticity of Blackness: A Non-conceptual Approach to the Study of Race and Racism in Fanon's and Merleau-Ponty's Phenomenology." *Human Architecture* 5 (2): 191–204.

Massumi, Brian. 2015. "The Supernormal Animal." In *The Nonhuman Turn*, edited by Richard R. Grusin, 1–18. Minneapolis: University of Minnesota Press.

McKibben, Bill. 1989. *The End of Nature.* New York: Anchor Books.

McWhorter, Ladelle. 1999. *Bodies and Pleasures: Foucault and the Politics of Sexual Normalization.* Bloomington: Indiana University Press.

Meacham, Darian. 2016. "How Low Can You Go? BioEnactivism, Cognitive Biology and Umwelt Ontology." *Humana.Mente Journal of Philosophical Studies* 9 (31): 73–95.

Meillassoux, Quentin. 2008. *After Finitude: An Essay on the Necessity of Contingency*. Translated by Ray Brassier. London: Continuum.

Meillassoux, Quentin, Ray Brassier, Ian Hamilton Grant, and Graham Harman. 2007. "Speculative Realism." *Collapse* 3 (November): 306–449.

Merchant, Carolyn. 1980. *The Death of Nature: Women, Ecology, and the Scientific Revolution*. San Francisco: Harper and Row.

———. 1996. *Earthcare: Women and the Environment*. New York: Routledge.

———. 2003. *Reinventing Eden: The Fate of Nature in Western Culture*. London: Routledge.

Merleau-Ponty, Maurice. 1962. *Phenomenology of Perception*. Translated by Colin Smith. London: Routledge.

———. 1963. *In Praise of Philosophy*. Translated by J. Wild and J. M. Edie. Evanston, IL: Northwestern University Press.

———. 1964a. *Sense and Non-Sense*. Translated by Hubert Dreyfus and Patricia Dreyfus. Evanston, IL: Northwestern University Press.

———. 1964b. *Signs*. Translated by Richard C. McLeary. Evanston, IL: Northwestern University Press.

———. 1964c. "The Primacy of Perception and Its Philosophical Consequences." Translated by J. M. Edie. In *The Primacy of Perception*, edited by J. M. Edie, 12–42. Evanston, IL: Northwestern University Press.

———. 1964d. "The Child's Relation with Others." Translated by W. Cobb. In *The Primacy of Perception*, edited by J. M. Edie, 96–156. Evanston, IL: Northwestern University Press.

———. 1964e. "Eye and Mind." Translated by Carleton Dallery. In *The Primacy of Perception*, edited by J. M. Edie, 159–92. Evanston, IL: Northwestern University Press.

———. 1965. *The Structure of Behaviour*. Translated by A. L. Fisher. London: Methuen.

———. 1968. *The Visible and the Invisible*. Translated by Alphonso Lingis. Evanston, IL: Northwestern University Press.

———. 2000. *Humanism and Terror: The Communist Problem*. Translated by John O'Neill. New Brunswick, NJ: Transaction Publishers.

———. 2003. *Nature: Course Notes from the Collège de France*. Translated by Robert Vallier. Evanston, IL: Northwestern University Press.

Micali, Stefano. 2013. "The Alteration of Embodiment in Melancholia." In *The Phenomenology of Embodied Subjectivity*, edited by Rasmus Thybo Jensen and Dermot Moran, 203–20. Dordrecht: Springer.

Miettinen, Timo. 2013. "The Body Politic: Husserl and the Embodied Community." In *The Phenomenology of Embodied Subjectivity*, edited by Rasmus Thybo Jensen and Dermot Moran, 329–46. Dordrecht: Springer.

Moore, Jason W. 2016. "The Rise of Cheap Nature." In *Anthropocene or Capitalocene: Nature, History and the Crisis of Capitalism*, edited by Jason W. Moore, 78–115. Oakland, CA: PM Press.

Morris, David. 2018. *Merleau-Ponty's Developmental Ontology*. Evanston, IL: Northwestern University Press.

Morton, Timothy. 2008. "Ecologocentrism: Unworking Animals." *Substance* 37 (3): 73–96.

———. 2010. "Queer Ecology." *PMLA* 125 (2): 273–82.

———. 2011. "Here Comes Everything: The Promise of Object-Oriented Ontology." *Qui Parle* 19 (2): 163–90.

———. 2012. "An Object-Oriented Defense of Poetry." *New Literary History* 43 (2): 205–24.

Naess, Arne. 1973. "The Shallow and the Deep, Long-Range Ecology Movement: A Summary." *Inquiry* 16 (1–4): 95–100.

Nussbaum, Martha. 2005. "Women and Cultural Universals." In *Feminist Theory: A Philosophical Anthology*, edited by Robin Andreasen and Ann Cudd, 302–24. Oxford: Blackwell.

Odling-Smee, John. 2009. "Niche Construction in Evolution, Ecosystems and Developmental Biology." In *Mapping the Future of Biology: Evolving Concepts and Theories*, edited by Anouk Barberousse, Michel Morange, and Thomas Pradeu, 69–91. Dordrecht: Springer.

Oksala, Johanna. 2006. "Female Freedom: Can the Lived Body Be Emancipated?" In *Feminist Interpretations of Maurice Merleau-Ponty*, edited by Dorothea Olkowski and Gail Weiss, 210–28. University Park: Pennsylvania State University Press.

Pavlov, Ivan P. 1927. *Conditioned Reflexes: An Investigation of the Physiological Activity of the Cerebral Cortex*. Translated and edited by G. V. Anrep. Oxford: Oxford University Press.

Persson, Ingmar, and Julian Savulescu. 2012. "Moral Enhancement, Freedom and the God Machine." *The Monist* 95 (3): 399–421.

Plato. 1965. *Timaeus and Critias*. Translated by Desmond Lee. Harmondsworth, UK: Penguin.

———. 1987. *The Republic*. Translated by Desmond Lee. Harmondsworth, UK: Penguin.

Plumptre, Andrew J., Alastair McNeilage, Jefferson S. Hall, and Liz Williamson. 2002. "The Current Status of Gorillas and Threats to Their Existence at the Beginning of a New Millennium." In *Gorilla Biology: A Multidisciplinary Perspective*, edited by Andrea B. Taylor and Michele L. Goldsmith, 414–31. Cambridge: Cambridge University Press.

Plumwood, Val. 1991. "Nature, Self, and Gender: Feminism, Environmental Philosophy, and the Critique of Rationalism." *Hypatia* 6 (1): 3–27.

———. 1993. *Feminism and the Mastery of Nature*. London: Routledge.

———. 2001. "Nature as Agency and the Prospects for a Progressive Naturalism." *Capitalism Nature Socialism* 12 (4): 3–32.

———. 2002. *Environmental Culture: The Ecological Crisis of Reason*. London: Routledge.

———. 2006. "The Concept of a Cultural Landscape: Nature, Culture and Agency in the Land." *Ethics and the Environment* 11 (2): 115–50.

Pollan, Michael. 2013. "The Intelligent Plant." *New Yorker*, September 23, 2013. https://www.michaelpollan.com/articles-archive/the-intelligent-plant.

Pope, Kyle, and Mark Hertsgaard. "Why Are the US News Media So Bad at Covering Climate Change?" *The Guardian*, April 22, 2019. https://www.theguardian.com/environment/2019/apr/22/why-is-the-us-news-media-so-bad-at-covering-climate-change.

Powell, James. 2019. "Scientists Reach 100% Consensus on Anthropogenic Global Warming." *Bulletin of Science, Technology & Society* 37 (4): 183–84.

Putnam, Hilary. 1981. *Reason, Truth and History*. Cambridge: Cambridge University Press.

Regan, Tom. 1983. *The Case for Animal Rights*. Berkeley: University of California Press.

Reynolds, Jack. 2004. *Merleau-Ponty and Derrida: Intertwining Embodiment and Alterity*. Athens: Ohio University Press.

———. 2016. "Merleau-Ponty's Gordian Knot: Transcendental Phenomenology, Science, and Naturalism." *Continental Philosophy Review* 50 (1): 81–104.

Richmond, Scott C. 2015. "Speculative Realism Is Speculative Aesthetics (Three New Books on Speculative Realism)." *Configurations* 23 (3): 399–403.

Rolston, Holmes, III. 1997. "Nature for Real: Is Nature a Social Construct?" In *The Philosophy of the Environment*, edited by T. D. J. Chappell, 38–64. Edinburgh: Edinburgh University Press.

Sachs, Carl B. 2017. "Discursive Intentionality as Embodied Coping: A Pragmatist Critique of Existential Phenomenology." In *Pragmatic Perspectives in Phenomenology*, edited by Ondrej Švec and Jakub apek, 87–102. London: Routledge.

———. 2014. *Intentionality and the Myths of the Given*. London: Pickering and Chatto.

Salleh, Ariel K. 1984. "Deeper Than Deep Ecology: The Eco-feminist Connection." *Environmental Ethics* 6 (4): 339–45.

Sartre, Jean-Paul. 1969. *Being and Nothingness*. Translated by Hazel E. Barnes. London: Routledge.

Saul, Jennifer. 2013. "Implicit Bias, Stereotype Threat and Women in Philosophy." In *Women in Philosophy: What Needs to Change?*, edited by Katrina Hutchison and Fiona Jenkins, 39–60. Oxford: Oxford University Press.

Scott, Jill W. 1992. "Experience." In *Feminists Theorize the Political*, edited by Judith Butler and Jill W. Scott, 22–40. London: Routledge.

Sen, Amartya. 1993. "India and the West." *New Republic*, June 7, 1993, 27–33.

Shaviro, Steven. 2014. *The Universe of Things*. Minneapolis: University of Minnesota Press.

Shiva, Vandana. 1991. "Staying Alive: Women, Ecology and Development." *Hypatia* 6 (1): 206–14.

Shusterman, Richard. 2005. "The Silent, Limping Body of Philosophy." In *The Cambridge Companion to Merleau-Ponty*, edited by Taylor Carman and Mark B. N. Hansen, 151–80. Cambridge: Cambridge University Press.

Silverman, Hugh. 1980. "Merleau-Ponty and the Interrogation of Language." *Research in Phenomenology* 10 (1): 122–41.

Singer, Peter. 1979. *Practical Ethics*. Cambridge: Cambridge University Press.

Sparrow, Tom. 2014. *The End of Phenomenology: Metaphysics and the New Realism*. Edinburgh: Edinburgh University Press.

Steel, Daniel. 2010. "Epistemic Values and the Argument from Inductive Risk." *Philosophy of Science* 77 (1): 14–34.

Steffen, Will, Paul Crutzen, and John R. McNeill. 2007. "The Anthropocene: Are Humans Now Overwhelming the Great Forces of Nature?" *Ambio* 36 (8): 614–21.

Steffen, Will, Johan Rockström, Katherine Richardson, Timothy M. Lenton, Carl Folke, Diana Liverman, Colin P. Summerhayes, Anthony D. Barnosky, Sarah E. Cornell, Michel Crucifix, Jonathan F. Donges, Ingo Fetzer, Steven J. Lade, Marten Scheffer, Ricarda Winkelmann, and Hans Joachim Schellnhuber. 2018.

"Trajectories of the Earth System in the Anthropocene." *Proceedings of the National Academy of Sciences of the United States of America* 115 (33): 8252–59.

Stenstad, Gail. 2011. *Thinking after Heidegger*. Madison: University of Wisconsin Press.

Steward, Helen. 2016. "Libertarianism as a Naturalistic Position." In *Free Will and Theism: Connections, Contingencies and Concerns*, edited by Kevin Timpe and Daniel Speak, 159–71. Oxford: Oxford University Press.

Stone, Alison. 2003. "The Sex of Nature: A Reinterpretation of Irigaray's Metaphysics and Political Thought." *Hypatia* 18 (3): 60–84.

———. 2008. "Unthought Nature: Reply to Penelope Deutscher and Mary Beth Mader." *Differences* 19 (3): 150–57.

———. 2015. "Irigaray's Ecological Phenomenology: Towards an Elemental Materialism." *Journal of the British Society for Phenomenology* 46 (2): 117–31.

Strawson, Galen. 2006. "Realistic Monism: Why Physicalism Entails Panpsychism." In *Consciousness and Its Place in Nature: Does Physicalism Entail Panpsychism?*, edited by Anthony Freeman, 3–31. Thorverton, UK: Imprint Academic.

Subcommission on Quaternary Stratigraphy. 2016. "Working Group on the Anthropocene." January 4, 2016. http://quaternary.stratigraphy.org/workinggroups/anthropocene/.

Sullivan, Shannon. 1997. "Domination and Dialogue in Merleau-Ponty's Phenomenology of Perception." *Hypatia* 12 (1): 1–19.

Thompson, Evan, and Mog Stapleton. 2009. "Making Sense of Sense-Making: Reflections on Enactive and Extended Mind Theories." *Topoi* 28 (1): 23–30.

Toadvine, Ted. 2008. "Phenomenology and Hyper-reflection." In *Merleau-Ponty: Key Concepts*, edited by Rosalyn Diprose and Jack Reynolds, 17–29. Stocksfield, UK: Acumen.

———. 2009. *Merleau-Ponty's Philosophy of Nature*. Evanston, IL: Northwestern University Press.

———. 2014. "The Elemental Past." *Research in Phenomenology* 44 (2): 262–79.

Trewavas, Anthony J., and František Baluška. 2011. "The Ubiquity of Consciousness." *EMBO Reports* 12 (12): 1221–25.

Tuana, Nancy. 2007. "Viscous Porosity: Witnessing Katrina." In *Material Feminisms*, edited by Stacy Alaimo and Susan Hekman, 188–213. Bloomington: Indiana University Press.

Twine, Richard. 2001. "Ma(r)king Essence—Ecofeminism and Embodiment." *Ethics and the Environment* 6 (2): 31–58.

Uexküll, Jakob von. 1957. "A Stroll through the Worlds of Animals and Men." In *Instinctive Behavior: The Development of a Modern Concept*, edited and translated by Claire H. Schiller, 5–80. New York: International Universities Press.

Urban, Petr. 2016. "Phenomenology and Interdisciplinarity." In *Kontexte des Leiblichen/Contexts of Corporality*, edited by Cathrin Nielsen, Karel Novotný, and Thomas Nenon, 459–80. Nordhausen, Germany: Verlag Traugott Bautz.

Vallier, Robert. 2009. "The Elemental Flesh: Nature, Life and Difference in Merleau-Ponty and Plato's Timaeus." In *Merleau-Ponty and the Possibilities of Philosophy*, edited by Bernard Flynn, Wayne J. Froman, and Robert Vallier, 129–53. Albany, NY: SUNY Press.

Varela, Francisco J., Evan Thompson, and Eleanor Rosch. 1991. *The Embodied Mind: Cognitive Science and Human Experience*. Cambridge, MA: MIT Press.

Vogel, Steven. 1996. *Against Nature*. Albany, NY: SUNY Press.

———. 2002. "Environmental Philosophy after the End of Nature." *Environmental Ethics* 24 (1): 23–39.

———. 2006. "The Silence of Nature." *Environmental Values* 15 (2): 145–71.

———. 2015. *Thinking Like a Mall: Environmental Philosophy after the End of Nature*. Cambridge, MA: MIT Press.

Wagemans, Johan, Jacob Feldman, Sergei Gepshtein, Ruth Kimchi, James R. Pomerantz, Peter A. van der Helm, and Cees van Leeuwen. 2012. "A Century of Gestalt Psychology in Visual Perception: II. Conceptual and Theoretical Foundations." *Psychological Bulletin* 138 (6): 1218–252.

Warren, Karen J. 1990. "The Power and Promise of Ecological Feminism." *Environmental Ethics* 12 (2): 125–46.

———. 2000. *Ecofeminist Philosophy: A Western Perspective on What It Is and Why It Matters*. Lanham, MD: Rowman and Littlefield.

Warren, Karen J., and Jim Cheney. 1991. "Ecological Feminism and Ecosystem Ecology." *Hypatia* 6 (1): 179–97.

Weate, Jeremy. 2001. "Fanon, Merleau-Ponty and the Difference of Phenomenology." In *Race,* edited by Robert Bernasconi, 169–83. Oxford: Blackwell.

Weiss, Gail. 1999. *Body Images: Embodiment as Intercorporeality*. London: Routledge.

Welsh, Talia. 2008. "The Developing Body: A Reading of Merleau-Ponty's Conception of Women in the Sorbonne Lectures." In *Intertwinings: Interdisciplinary Encounters with Merleau-Ponty*, edited by Gail Weiss, 45–59. Albany, NY: SUNY Press.

Westling, Louise. 2014. *The Logos of the Living World: Merleau-Ponty, Animals, and Language*. New York: Fordham University Press.

Wilson, Edward O. 1998. *Consilience: The Unity of Knowledge*. New York: Random House.

Wollstonecraft, Mary. 1995. *A Vindication of the Rights of Men and a Vindication of the Rights of Women*. Cambridge: Cambridge University Press.

Young, Iris Marion. 1989. "Throwing Like a Girl." In *The Thinking Muse: Feminism and Modern French Philosophy*, edited by Jeffner Allen and Iris Marion Young, 51–70. Bloomington: Indiana University Press.

Zahavi, Dan. 2016. "The End of What? Phenomenology vs. Speculative Realism." *International Journal of Philosophical Studies* 24 (3): 289–309.

Zhang Jennifer. 2020. "Coronavirus Response Shows the World Is Not Ready for Climate-Induced Pandemics." *State of the Planet* (blog), Earth Institute, Columbia University, February 24, 2020. https://blogs.ei.columbia.edu/2020/02/24/coronavirus-climate-induced-pandemic.

Žižek, Slavoj. 2008. *Violence*. London: Profile Books.

INDEX